# SpringerBriefs in Energy

SpringerBriefs in Energy presents concise summaries of cutting-edge research and practical applications in all aspects of Energy. Featuring compact volumes of 50 to 125 pages, the series covers a range of content from professional to academic. Typical topics might include:

- A snapshot of a hot or emerging topic
- A contextual literature review
- A timely report of state-of-the art analytical techniques
- An in-depth case study
- A presentation of core concepts that students must understand in order to make independent contributions.

Briefs allow authors to present their ideas and readers to absorb them with minimal time investment.

Briefs will be published as part of Springer's eBook collection, with millions of users worldwide. In addition, Briefs will be available for individual print and electronic purchase. Briefs are characterized by fast, global electronic dissemination, standard publishing contracts, easy-to-use manuscript preparation and formatting guidelines, and expedited production schedules. We aim for publication 8-12 weeks after acceptance.

Both solicited and unsolicited manuscripts are considered for publication in this series. Briefs can also arise from the scale up of a planned chapter. Instead of simply contributing to an edited volume, the author gets an authored book with the space necessary to provide more data, fundamentals and background on the subject, methodology, future outlook, etc.

SpringerBriefs in Energy contains a distinct subseries focusing on Energy Analysis and edited by Charles Hall, State University of New York. Books for this subseries will emphasize quantitative accounting of energy use and availability, including the potential and limitations of new technologies in terms of energy returned on energy invested.

More information about this series at http://www.springer.com/series/8903

Rafael J. Bergillos · Cristobal Rodriguez-Delgado ·
Gregorio Iglesias

# Ocean Energy and Coastal Protection

## A Novel Strategy for Coastal Management Under Climate Change

 Springer

Rafael J. Bergillos🆔
Hydraulic Engineering Area
Department of Agronomy
University of Córdoba
Córdoba, Spain

Gregorio Iglesias🆔
MaREI, Environmental Research
Institute and School of Engineering
University College Cork
Cork, Ireland

Cristobal Rodriguez-Delgado🆔
School of Engineering
University of Plymouth
Plymouth, UK

ISSN 2191-5520    ISSN 2191-5539   (electronic)
SpringerBriefs in Energy
ISBN 978-3-030-31317-3    ISBN 978-3-030-31318-0   (eBook)
https://doi.org/10.1007/978-3-030-31318-0

This Springer imprint is published by the registered company Springer Nature Switzerland AG
The registered company address is: Gewerbestrasse 11, 6330 Cham, Switzerland

# Preface

This Springer Briefs volume on ocean energy and coastal protection contributes to the development of a novel strategy for coastal management in a climate change context: the implementation of wave energy converter farms for the dual function of renewable energy generation and coastal protection against erosion and flooding.

The book comprises the development and application of artificial intelligence to optimize wave farm location and layout, the assessment of the influence of the wave energy converter geometry on coastal protection against erosion and flooding, and the analysis of how the performance of wave farms as coastal defence elements is affected by climate change and sea-level rise. The concepts, methods, results and conclusions presented in this book are helpful to students, researchers, academics, engineers, designers, stakeholders and managers across the world interested in wave energy, coastal engineering and coastal management.

This work has been carried out in the framework of the research grants WAVEIMPACT (PCIG-13-GA-2013-618556, European Commission, Marie Curie fellowship, fellow GI) and ICE (Intelligent Community Energy, European Commission, Contract no. 5025). RB was partly funded by the Spanish Ministry of Science, Innovation and Universities (*Programa Juan de la Cierva 2017*, FJCI-2017-31781). CRD was partly funded by the University of Plymouth (UK).

Wave, sea-level rise, bathymetric and DEM data were provided by *Puertos del Estado* (Spain), *Universität Hamburg* (Germany), *Ministerio de Agricultura, Pesca y Alimentación* (Spain) and *Instituto Geográfico Nacional* (Spain), respectively. RB and CRD are grateful to Prof. Miguel Ortega-Sánchez for his training during their student stages. The authors also thank James Allen for his support with the laboratory experiments.

Córdoba, Spain                                          Rafael J. Bergillos
Plymouth, UK                              Cristobal Rodriguez-Delgado
Cork, Ireland                                             Gregorio Iglesias
September 2019

# Contents

# Chapter 1
# Introduction

**Abstract** Among the challenges that managers and policy-makers should confront in the coming years, two of the most relevant are the development and use of carbon-free energy sources and the adaptation to the consequences of climate change, including sea-level rise. Wave energy is one of the most promising renewable energy sources due its great potential and its low environmental impacts, including visual pollution. This chapter presents a summary of both the overview in ocean energy and the current state of the art in wave energy. Then, the main objective of the book and the selected study site are specified. Finally, it is indicated the structure of the book, detailing the main purposes of each chapter.

## 1.1 An Overview of Ocean Energy

The mitigation of the consequences of climate change and the use of renewable energy sources are two of the main challenges that society will face in the next decades, partly due to the finite availability of fossil fuels, their negative environmental impacts and their high costs [5, 89]. Accordingly, the main worldwide governments are focusing their policies on the development of carbon-free energy [14, 31, 40, 42, 45, 48, 57, 82, 87].

Among the renewable energy typologies, ocean energy sources are attractive and promising due to their comparatively extensive worldwide availability, high power density and low negative effects on the environment [35, 36, 44, 78, 83]. In particular, the wave energy potential across the world is in the order of 17 TW h/y [70]. Despite these facts, the degree of development of this source is still limited compared to other carbon-free energy sources such as wind energy, hydroelectric or biomass energy [43, 92]. This has triggered increasing research efforts and progresses focused on wave energy. A review of the main studies in relation to wave energy over the past decade is detailed in the following section.

© The Author(s), under exclusive license to Springer Nature Switzerland AG 2020
R. J. Bergillos et al., *Ocean Energy and Coastal Protection*,
SpringerBriefs in Energy, https://doi.org/10.1007/978-3-030-31318-0_1

## 1.2 State of the Art in Wave Energy

Most of the works in relation to wave energy so far have been focused on the assessment of the potential at different locations [27–29, 33, 50, 53–55, 66–69, 80, 90, 91, 93, 94, 98], development of wave energy converters and technologies [13, 26, 30, 34, 37–39, 41, 46, 49, 56, 58–65, 71–73, 76, 81, 88, 95–98, 100–102] and the economic viability of this energy source [8, 10, 11, 32, 47].

Recent studies have also explored the combined implementation and exploitation of wave energy and offshore wind energy [6, 7, 9, 12, 79, 99] and the dual function of wave energy converter farms as carbon-free energy generators and coastal defence elements against erosion [1–4, 15, 74, 75, 84–86].

## 1.3 Objective and Outline of the Book

This book provides new insights into the development of a novel strategy for coastal management in a climate change context: the implementation of wave energy converter farms for the dual function of renewable energy generation and coastal protection against erosion and flooding. For this purpose, WaveCat devices, which are floating overtopping wave energy converters consisting of twin hulls joined by their stern, are considered [51, 52].

The methods, tools and strategies presented in the work are based on a case study in southern Iberian Peninsula (Playa Granada). This deltaic beach has been experiencing strong erosion and flooding issues in the past two decades, which were exacerbated by human interventions in the Guadalfeo River basin [16–25, 77]. However, the management strategy and the methodology applied to optimize the design of the strategy are feasibly extensible to other coasts across the world to mitigate the consequences of climate change.

The book is structured as follows. Chapter 2 develops and applies artificial intelligence to optimize the wave farm location and layout for coastal protection. The influence of the wave energy converter configuration on coastal defence against dry beach erosion and coastal flooding is addressed in Chaps. 3 and 4, respectively. Finally, Chaps. 5 and 6 analyse how sea-level rise affect the performance of wave farms for coastal protection against erosion and flooding, respectively.

## References

1. Abanades J, Flor-Blanco G, Flor G, Iglesias G (2018) Dual wave farms for energy production and coastal protection. Ocean Coast Manag 160:18–29. https://doi.org/10.1016/j.ocecoaman.2018.03.038
2. Abanades J, Greaves D, Iglesias G (2014) Coastal defence through wave farms. Coast Eng 91:299–307

3. Abanades J, Greaves D, Iglesias G (2014) Wave farm impact on the beach profile: a case study. Coast Eng 86:36–44
4. Abanades J, Greaves D, Iglesias G (2015) Coastal defence using wave farms: the role of farm-to-coast distance. Renew Energy 75:572–582
5. Asif M, Muneer T (2007) Energy supply, its demand and security issues for developed and emerging economies. Renew Sustain Energy Rev 11(7):1388–1413
6. Astariz S, Abanades J, Perez-Collazo C, Iglesias G (2015) Improving wind farm accessibility for operation and maintenance through a co-located wave farm: influence of layout and wave climate. Energy Convers Manag 95:229–241
7. Astariz S, Iglesias G (2015) Enhancing wave energy competitiveness through co-located wind and wave energy farms. a review on the shadow effect. Energies 8(7):7344–7366
8. Astariz S, Iglesias G (2015) The economics of wave energy: a review. Renew Sustain Energy Rev 45:397–408
9. Astariz S, Iglesias G (2016) Output power smoothing and reduced downtime period by combined wind and wave energy farms. Energy 97:69–81. https://doi.org/10.1016/j.energy.2015.12.108
10. Astariz S, Iglesias G (2016) Wave energy vs. other energy sources: a reassessment of the economics. Int J Green Energy 13(7):747–755. https://doi.org/10.1080/15435075.2014.963587
11. Astariz S, Vazquez A, Iglesias G (2015) Evaluation and comparison of the levelized cost of tidal, wave, and offshore wind energy. J Renew Sustain Energy 7(5):053112. https://doi.org/10.1063/1.4932154
12. Azzellino A, Ferrante V, Kofoed JP, Lanfredi C, Vicinanza D (2013) Optimal siting of offshore wind-power combined with wave energy through a marine spatial planning approach. Int J Mar Energy 3:e11–e25
13. Barambones O, Cortajarena JA, de Durana JMG, Alkorta P (2018) A real time sliding mode control for a wave energy converter based on a wells turbine. Ocean Eng 163:275–287
14. Bekun FV, Alola AA, Sarkodie SA (2018) Toward a sustainable environment: nexus between CO2 emissions, resource rent, renewable and nonrenewable energy in 16-EU countries. Sci Total Environ 657:1023–1029
15. Bergillos RJ, Lopez-Ruiz A, Medina-Lopez E, Monino A, Ortega-Sanchez M (2018) The role of wave energy converter farms on coastal protection in eroding deltas, Guadalfeo, southern Spain. J Clean Prod 171:356–367
16. Bergillos RJ, López-Ruiz A, Ortega-Sánchez M, Masselink G, Losada MA (2016) Implications of delta retreat on wave propagation and longshore sediment transport-Guadalfeo case study (southern Spain). Mar Geol 382:1–16
17. Bergillos RJ, López-Ruiz A, Principal-Gómez D, Ortega-Sánchez M (2018) An integrated methodology to forecast the efficiency of nourishment strategies in eroding deltas. Sci Total Environ 613:1175–1184
18. Bergillos RJ, Masselink G, McCall RT, Ortega-Sánchez M (2016) Modelling overwash vulnerability along mixed sand-gravel coasts with XBeach-G: case study of Playa Granada, southern Spain. In: Proceedings of the coastal engineering, vol. 1(35), p. 13
19. Bergillos RJ, Masselink G, Ortega-Sánchez M (2017) Coupling cross-shore and longshore sediment transport to model storm response along a mixed sand-gravel coast under varying wave directions. Coast Eng 129:93–104
20. Bergillos RJ, Ortega-Sánchez M (2017) Assessing and mitigating the landscape effects of river damming on the Guadalfeo River delta, southern Spain. Landsc Urban Plan 165:117–129
21. Bergillos RJ, Ortega-Sánchez M, Losada MA (2015) Foreshore evolution of a mixed sand and gravel beach: the case of Playa Granada (southern Spain). In: Proceedings of the 8th coastal sediments, World Scientific, Singapore
22. Bergillos RJ, Ortega-Sánchez M, Masselink G, Losada MA (2016) Morpho-sedimentary dynamics of a micro-tidal mixed sand and gravel beach, Playa Granada, southern Spain. Mar Geol 379:28–38
23. Bergillos RJ, Rodríguez-Delgado C, López-Ruiz A, Millares A, Ortega-Sánchez M, Losada MA (2015) Recent human-induced coastal changes in the Guadalfeo river deltaic system

(southern Spain). In: Proceedings of the 36th IAHR-international association for hydro-environment engineering and research world congress

24. Bergillos RJ, Rodríguez-Delgado C, Millares A, Ortega-Sánchez M, Losada MA (2016) Impact of river regulation on a Mediterranean delta: assessment of managed versus unmanaged scenarios. Water Resour Res 52(7):5132–5148

25. Bergillos RJ, Rodríguez-Delgado C, Ortega-Sánchez M (2017) Advances in management tools for modeling artificial nourishments in mixed beaches. J Mar Syst 172:1–13

26. Buccino M, Stagonas D, Vicinanza D (2015) Development of a composite sea wall wave energy converter system. Renew Energy 81:509–522

27. Carballo R, Arean N, Álvarez M, López I, Castro A, López M, Iglesias G (2018) Wave farm planning through high-resolution resource and performance characterization. Renew Energy 135:1097–1107

28. Carballo R, Sánchez M, Ramos V, Fraguela J, Iglesias G (2015) The intra-annual variability in the performance of wave energy converters: a comparative study in N Galicia (Spain). Energy 82:138–146

29. Carballo R, Sánchez M, Ramos V, Taveira-Pinto F, Iglesias G (2014) A high resolution geospatial database for wave energy exploitation. Energy 68:572–583

30. Chao Z, Yage Y, Aiju C (2018) Hydrodynamics research of a two-body articulated wave energy device. Ocean Eng 148:202–210

31. Commission European (2007) A european strategic energy technology plan (set-plan): towards a low carbon future. Commission of the European Communities, Brussels

32. Contestabile P, Di Lauro E, Buccino M, Vicinanza D (2017) Economic assessment of over-topping breakwater for energy conversion (OBREC): a case study in western Australia. Sustainability 9(1):51. https://doi.org/10.3390/su9010051

33. Contestabile P, Ferrante V, Vicinanza D (2015) Wave energy resource along the coast of Santa Catarina (Brazil). Energies 8(12):14219–14243

34. Contestabile P, Iuppa C, Di Lauro E, Cavallaro L, Andersen TL, Vicinanza D (2017) Wave loadings acting on innovative rubble mound breakwater for overtopping wave energy conversion. Coast Eng 122:60–74

35. Cornett AM (2008) A global wave energy resource assessment. In: The 18th international offshore and polar engineering conference. International Society of Offshore and Polar Engineers, Mountain View

36. Cruz J (2008) Ocean wave energy: current status and future perspectives. Springer Science & Business Media, Berlin

37. de O. Falcão, AF (2007) Modelling and control of oscillating-body wave energy converters with hydraulic power take-off and gas accumulator. Ocean Eng 34(14):2021–2032

38. Day A, Babarit A, Fontaine A, He YP, Kraskowski M, Murai M, Penesis I, Salvatore F, Shin HK (2015) Hydrodynamic modelling of marine renewable energy devices: a state of the art review. Ocean Eng 108:46–69

39. Do HT, Dang TD, Ahn KK (2018) A multi-point-absorber wave-energy converter for the stabilization of output power. Ocean Eng 161:337–349

40. Dong K, Sun R, Dong X (2018) CO 2 emissions, natural gas and renewables, economic growth: assessing the evidence from China. Sci Total Environ 640:293–302

41. Elhanafi A, Macfarlane G, Fleming A, Leong Z (2017) Experimental and numerical investigations on the hydrodynamic performance of a floating-moored oscillating water column wave energy converter. Appl Energy 205:369–390

42. European Commission (2009) Renewable energy directive 2009/28/EC . European Union, Brussels

43. Eurostat (2016) Renewable energy statistics. European Union, Brussels

44. Falnes J (2007) A review of wave-energy extraction. Mar Struct 20(4):185–201

45. Feng T, Zhou W, Wu S, Niu Z, Cheng P, Xiong X, Li G (2018) Simulations of summertime fossil fuel CO 2 in the Guanzhong basin China. Sci Total Environ 624:1163–1170

46. Fernandez H, Iglesias G, Carballo R, Castro A, Fraguela J, Taveira-Pinto F, Sanchez M (2012) The new wave energy converter WaveCat: concept and laboratory tests. Mar Struct 29:58–70

47. Frost C, Findlay D, Macpherson E, Sayer P, Johanning L (2018) A model to map levelised cost of energy for wave energy projects. Ocean Eng 149:438–451
48. Gaete-Morales C, Gallego-Schmid A, Stamford L, Azapagic A (2018) Assessing the environmental sustainability of electricity generation in Chile. Sci Total Environ 636:1155–1170
49. Halder P, Mohamed MH, Samad A (2018) Wave energy conversion: design and shape optimization. Ocean Eng 150:337–351
50. Iglesias G, Carballo R (2011) Choosing the site for the first wave farm in a region: a case study in the Galician Southwest (Spain). Energy 36:5525–5531
51. Iglesias G, Carballo R, Castro A, Fraga B (2009) Development and design of the WaveCat™ energy converter. In: Coastal engineering 2008, vol. 5. World Scientific, Singapore, pp. 3970–3982
52. Iglesias G, Fernándes H, Carballo R, Castro A, Taveira-Pinto F (2011) The wavecat-development of a new wave energy converter. In: World renewable energy congress-Sweden, vol 57. Linköping University Electronic Press, Linköping, Sweden. pp 2151–2158. Accessed 8–13 May 2011
53. Iglesias G, López M, Carballo R, Castro A, Fraguela JA, Frigaard P (2009) Wave energy potential in Galicia (NW Spain). Renew Energy 34(11):2323–2333
54. Iuppa C, Cavallaro L, Foti E, Vicinanza D (2015) Potential wave energy production by different wave energy converters around Sicily. J Renew Sustain Energy 7(6):061701. https://doi.org/10.1063/1.4936397
55. Iuppa C, Cavallaro L, Vicinanza D, Foti E (2015) Investigation of suitable sites for wave energy converters around Sicily (Italy). Ocean Sci Discuss 12(1):315–354
56. Kolios A, Di Maio LF, Wang L, Cui L, Sheng Q (2018) Reliability assessment of point-absorber wave energy converters. Ocean Eng 163:40–50
57. Lin B, Zhu J (2019) The role of renewable energy technological innovation on climate change: empirical evidence from China. Sci Total Environ 659:1505–1512
58. López I, Castro A, Iglesias G (2015) Hydrodynamic performance of an oscillating water column wave energy converter by means of particle imaging velocimetry. Energy 83:89–103. https://doi.org/10.1016/j.energy.2015.01.119
59. López I, Iglesias G (2014) Efficiency of OWC wave energy converters: a virtual laboratory. Appl Ocean Res 44:63–70
60. López I, Pereiras B, Castro F, Iglesias G (2014) Optimisation of turbine-induced damping for an OWC wave energy converter using a RANS-VOF numerical model. Appl Energy 127:105–114
61. López I, Pereiras B, Castro F, Iglesias G (2015) Performance of OWC wave energy converters: influence of turbine damping and tidal variability. Int J Energy Res 39(4), 472–483 . ER-13-4164.R2
62. López I, Pereiras B, Castro F, Iglesias G (2016) Holistic performance analysis and turbine-induced damping for an owc wave energy converter. Renew Energy 85:1155–1163
63. López M, Ramos V, Rosa-Santos P, Taveira-Pinto F (2018) Effects of the PTO inclination on the performance of the CECO wave energy converter. Mar Struct 61:452–466
64. López M, Taveira-Pinto F, Rosa-Santos P (2017) Influence of the power take-off characteristics on the performance of ceco wave energy converter. Energy 120:686–697
65. López M, Taveira-Pinto F, Rosa-Santos P (2017) Numerical modelling of the ceco wave energy converter. Renew Energy 113:202–210
66. López M, Veigas M, Iglesias G (2015) On the wave energy resource of Peru. Energy Convers Manag 90:34–40
67. López-Ruiz A, Bergillos RJ, Lira-Loarca A, Ortega-Sánchez M (2018) A methodology for the long-term simulation and uncertainty analysis of the operational lifetime performance of wave energy converter arrays. Energy 153:126–135
68. López-Ruiz A, Bergillos RJ, Ortega-Sánchez M (2016) The importance of wave climate forecasting on the decision-making process for nearshore wave energy exploitation. Appl Energy 182:191–203

69. López-Ruiz A, Bergillos RJ, Raffo-Caballero JM, Ortega-Sánchez M (2018) Towards an optimum design of wave energy converter arrays through an integrated approach of life cycle performance and operational capacity. Appl Energy 209:20–32

70. Lund H (2007) Renewable energy strategies for sustainable development. Energy 32(6):912–919

71. Margheritini L, Vicinanza D, Frigaard P (2009) SSG wave energy converter: design, reliability and hydraulic performance of an innovative overtopping device. Renew Energy 34(5):1371–1380

72. Medina-López E, Bergillos R, Moñino A, Clavero M, Ortega-Sánchez M (2017) Effects of seabed morphology on oscillating water column wave energy converters. Energy 135:659–673

73. Medina-López E, Moñino A, Bergillos R, Clavero M, Ortega-Sánchez M (2019) Oscillating water column performance under the influence of storm development. Energy 166:765–774

74. Mendoza E, Silva R, Zanuttigh B, Angelelli E, Andersen TL, Martinelli L, Nørgaard JQH, Ruol P (2014) Beach response to wave energy converter farms acting as coastal defence. Coast Eng 87:97–111

75. Millar D, Smith H, Reeve D (2007) Modelling analysis of the sensitivity of shoreline change to a wave farm. Ocean Eng 34:884–901

76. Moñino A, Medina-López E, Bergillos RJ, Clavero M, Borthwick A, Ortega-Sánchez M (2018) Thermodynamics and morphodynamics in wave energy. Springer, Berlin

77. Ortega-Sánchez M, Bergillos RJ, López-Ruiz A, Losada MA (2017) Morphodynamics of Mediterranean mixed sand and gravel coasts. Springer, Berlin

78. Panwar N, Kaushik S, Kothari S (2011) Role of renewable energy sources in environmental protection: a review. Renew Sustain Energy Rev 15(3):1513–1524

79. Pérez-Collazo C, Greaves D, Iglesias G (2015) A review of combined wave and offshore wind energy. Renew Sustain Energy Rev 42:141–153. https://doi.org/10.1016/j.rser.2014.09.032

80. Prieto LF, Rodríguez GR, Rodríguez JS (2019) Wave energy to power a desalination plant in the north of Gran Canaria Island: wave resource, socioeconomic and environmental assessment. J Environ Manag 231:546–551

81. Ramos V, López M, Taveira-Pinto F, Rosa-Santos P (2018) Performance assessment of the ceco wave energy converter: water depth influence. Renew Energy 117:341–356

82. Rana R, Ingrao C, Lombardi M, Tricase C (2016) Greenhouse gas emissions of an agro-biogas energy system: estimation under the renewable energy directive. Sci Total Environ 550:1182–1195

83. Rinaldi G, Thies P, Walker R, Johanning L (2017) A decision support model to optimise the operation and maintenance strategies of an offshore renewable energy farm. Ocean Eng 145:250–262

84. Rodriguez-Delgado C, Bergillos RJ, Iglesias G (2019) Dual wave energy converter farms and coastline dynamics: the role of inter-device spacing. Sci Total Environ 646:1241–1252

85. Rodriguez-Delgado C, Bergillos RJ, Ortega-Sánchez M, Iglesias G (2018) Protection of gravel-dominated coasts through wave farms: layout and shoreline evolution. Sci Total Environ 636:1541–1552

86. Rodriguez-Delgado C, Bergillos RJ, Ortega-Sánchez M, Iglesias G (2018) Wave farm effects on the coast: the alongshore position. Sci Total Environ 640:1176–1186

87. Sarkodie SA, Adams S (2018) Renewable energy, nuclear energy, and environmental pollution: accounting for political institutional quality in South Africa. Sci Total Environ 643:1590–1601

88. Sergiienko N, Rafiee A, Cazzolato B, Ding B, Arjomandi M (2018) Feasibility study of the three-tether axisymmetric wave energy converter. Ocean Eng 150:221–233

89. Shafiee S, Topal E (2009) When will fossil fuel reserves be diminished? Energy policy 37(1):181–189

90. Silva D, Bento AR, Martinho P, Soares CG (2015) High resolution local wave energy modelling in the Iberian Peninsula. Energy 91:1099–1112

91. Veigas M, López M, Iglesias G (2014) Assessing the optimal location for a shoreline wave energy converter. Appl Energy 132:404–411

92. U.S. Energy Information Administration et al (2011) Annual energy outlook 2011: with projections to 2035. Government Printing Office, Washington, D.C

93. Veigas M, López M, Romillo P, Carballo R, Castro A, Iglesias G (2015) A proposed wave farm on the Galician coast. Energy Convers Manag 99:102–111

94. Vicinanza D, Contestabile P, Ferrante V (2013) Wave energy potential in the north-west of Sardinia (Italy). Renew Energy 50:506–521

95. Vicinanza D, Contestabile P, Nørgaard JQH, Andersen TL (2014) Innovative rubble mound breakwaters for overtopping wave energy conversion. Coast Eng 88:154–170

96. Vicinanza D, Margheritini L, Kofoed JP, Buccino M (2012) The SSG wave energy converter: performance, status and recent developments. Energies 5:193–226

97. Vicinanza D, Nørgaard JH, Contestabile P, Andersen TL (2013) Wave loadings acting on overtopping breakwater for energy conversion. J Coast Res 65(sp2):1669–1674

98. Viviano A, Naty S, Foti E, Bruce T, Allsop W, Vicinanza D (2016) Large-scale experiments on the behaviour of a generalised oscillating water column under random waves. Renew Energy 99:875–887

99. Wang Z, Duan C, Dong S (2018) Long-term wind and wave energy resource assessment in the South China sea based on 30-year hindcast data. Ocean Eng 163:58–75

100. Wu B, Li M, Wu R, Chen T, Zhang Y, Ye Y (2018) BBDB wave energy conversion technology and perspective in China. Ocean Eng 169:281–291

101. Yang Y, Diaz I, Morales M (2018) A vertical-axis unidirectional rotor for wave energy conversion. Ocean Eng 160:224–230

102. Zheng S, Zhang Y (2018) Analytical study on wave power extraction from a hybrid wave energy converter. Ocean Eng 165:252–263

# Chapter 2
# Optimization of Wave Farm Location and Layout for Coastal Protection

**Abstract** In this chapter, a new methodology to manage coastal protection by means of wave farms is proposed. Artificial intelligence tools, more specifically artificial neural networks (ANNs), were used to assess dry beach surface differences between the no wave farm situation and different wave farm project scenarios. A number of alongshore locations and layouts—represented as the number of rows and the spacing between devices—formed the wave farm scenarios and the influence of the wave climate—significant wave height and mean wave direction—was also taken into account. The selected study site was Playa Granada (southern Iberian Peninsula), a beach with important erosion problems. The datasets used for training and testing the ANN were obtained by means of a suite of numerical models including a third generation wave propagation model, a sediment transport formulation and a shoreline response equation. In order to obtain the ANN which provides the best fit to the data, a comparative study involving more than forty different architectures formed by one and two hidden layers and trained by means of two training algorithm was carried out. The [5-10-1] architecture obtained the best results with a correlation coefficient and RMSE of 0.9489 and 4.22 $m^2$, respectively. Once the best architecture was found, the ANN was applied to the study site in order to obtain the optimum location and layout for a wave farm project and the results indicate that dry beach surface could increase up to 5400.18 $m^2$ per year. These results show that ANNs can be useful for managers to optimize the design of wave farms for coastal protection.

## 2.1 Objective

The objective of this second chapter is to develop a method to design a wave farm in order to achieve its optimum performance in terms of coastal protection. To do that, artificial intelligence and, more precisely, artificial neural networks (ANN) was used. By means of ANN, the layout and location of the wave farm were optimized given certain wave conditions.

The outline of the chapter is as follows: Sect. 2.2 shows the characteristics of the study area (Playa Granada, Southern Iberian Peninsula). Section 2.3 describes the methodology used to develop the ANN and to obtain the data for training and

© The Author(s), under exclusive license to Springer Nature Switzerland AG 2020    9
R. J. Bergillos et al., *Ocean Energy and Coastal Protection*,
SpringerBriefs in Energy, https://doi.org/10.1007/978-3-030-31318-0_2

validating it, which includes wave propagation, sediment transport and shoreline position numerical models. Finally, the results obtained are depicted in Sect. 2.4 and the conclusions drawn in Sect. 2.5.

## 2.2  Study Area

The developed ANN was applied to a study area in order to highlight its efficiency to optimize the coastal protection provided by a wave farm. The selected study area was Playa Granada, a gravel dominated beach in Southern Iberian Peninsula, facing the Mediterranean sea (Fig. 2.1). The west and east limits of this stretch of coast are the Guadalfeo River mouth and Punta del Santo (a shoreline horn), respectively. In 2004, the Guadalfeo River basin run off was regulated which, over the last decade, has induced erosion problems and severe coastline retreat [2, 5, 7, 27], as shown in Fig. 2.2. During the next decade to the river regulation, more than 4000 m$^2$/year of dry beach area were lost and the maximum retreat of the shoreline was 87 m [3, 11].

In order to solve these problems, hard-engineering solutions such as artificial beach nourishments have been carried out in the region. However, this solutions have not been successful in a long-term [6, 9, 10, 12]. Because of that, it would be

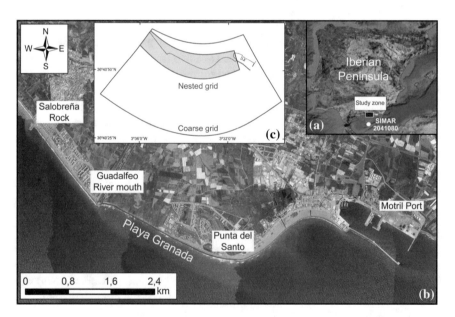

**Fig. 2.1** **a** Location of the study area in southern Spain, **b** plan view of the deltaic coast, indicating the stretch of beach studied (Playa Granada), **c** boundaries of the computational grids used by the wave propagation model (*Source* [31]. Reproduced with permission of Elsevier)

**Fig. 2.2** Study area: **a** erosion and coastal flooding, **b** replenishment works (*Source* [31]. Reproduced with permission of Elsevier)

necessary to deploy new management strategies, focused on the long-term, in order to slow down the erosion at Playa Granada.

In addition, the ability of WECs to produce carbon-free energy with the short periods and relatively low wave heights of the Mediterranean has been proved in recent works [20, 22–24]. Reference [25] assessed the wave energy potential specifically in this region of the Alboran sea. Moreover, the numerical models needed to study the sediment transport patterns and the shoreline movements in order to investigate the erosion problems in Playa Granada have been already validated in the study area, based on comparison with data from field campaigns [4, 12]. These characteristics make Playa Granada an ideal location to study the possibility of using hybrid wave farms, for energy extraction and coast protection.

## 2.3  Methods

Figure 2.3 shows the methodology followed in this chapter. Using wave propagation, longshore sediment transport and shoreline position numerical models the dataset for the training and validating process of the artificial neural networks was generated. Then, the best configuration of the ANN to fit the dataset generated was obtained testing different architectures and training algorithms. Finally, the wave farm which provide the best coastal protection performance was found using the final artificial neural network configuration.

### 2.3.1  Artificial Neural Networks

Artificial neural networks (ANNs) are artificial intelligence tools formed by a group of elements (*neurons*) which are able to work in parallel [15, 16]. This systems are inspired by biological brains—hence its name—, emulating the connections of neurons inside them. They are able to learn from experience through the process of training. Artificial neural networks are able to learn from data and perform complicated task with no prior of the dataset. Multilayer feed forward ANNs, which we are dealing with in this work, are usually applied to non-linear regression and function approximation.

The connection between the artificial neurons are formed by weights and biases. In order to adjust them, before the training, the initial dataset is divided in two groups, the training and testing dataset. Each of this datasets are formed by input and

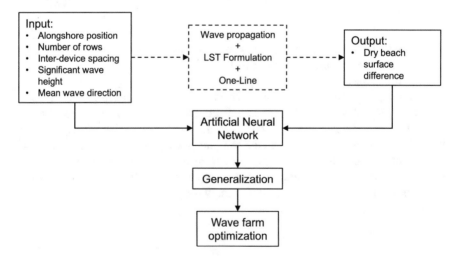

**Fig. 2.3** Flow chart of the proposed methodology (*Source* [31]. Reproduced with permission of Elsevier)

**Fig. 2.4** Schematic concept of a multilayer feedforward artificial neural network (*Source* [31]. Reproduced with permission of Elsevier)

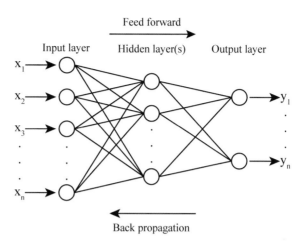

their corresponding target vectors. During the process of training the input vectors are presented to the ANN, which propagates them through the different layers to produce output vectors. Then, during the process of testing, the output produced by the ANN and the original target vector are compared and the error between them is assessed (normally through the MSE). Finally, the weight and biases are changed in order to minimise the obtained error. This process is repeated until a specified target is achieved, whether a desired value of MSE or a specific number of iterations.

In this work, multilayer feedforward ANNs are used. This kind of ANNs are formed by an input layer, a number of hidden layer and finally, an output layer (Fig. 2.4). Inside each layer, the output of each neuron is produced by means of a derivable function. Different functions have been used with different purposes in the literature but, for this work, the sigmoidal transfer function is applied in the neurons of the hidden layer. Its expression is as follows:

$$y = 2\left(1 + e^{-2x}\right)^{-1} - 1. \tag{2.1}$$

Finally, for the output layer a linear function was applied. It is defined as:

$$y = x. \tag{2.2}$$

In this work, two different training algorithms have been considered: Levenberg–Marquardt [14] and Bayesian regularization [26]. These algorithms have been chosen because their proven efficiency to solve coastal engineering problems. Reference [21] applied them to assess the efficiency of a new developed WEC, in addition, they are usually applied to wave forecasting [19] and also have been used successfully to predict the evolution of the shoreline position [18].

Their differences consist in the different approaching they have to the training process. A faster training is achieved using Levenberg–Marquardt algorithm. In this case, the aforementioned test dataset is divided in two sub-groups: proper test and validation. After each iteration, the ANN is applied to the validation dataset and its MSE is obtained. The training stops when this validation MSE grow up between two iterations. The Bayesian regularization algorithm does not need a validation dataset and the basis to stop the training is weight minimization. This algorithm usually achieves good results for noisy of difficult data.

### 2.3.2   Numerical Models

The first step to develop the ANN is the generation of the dataset. In this case, a group of numerical models were used to obtain the training and testing dataset. First, sea states are propagated from deep water to the nearshore by means of a third-generation wave propagation model. Then, sediment transport patterns are obtained by means of an LST formulation and finally, shoreline position is assessed using a one-line model. This numerical scheme have been applied successfully to the study site in previous works [12, 29, 30].

To represent the wave climate of Playa Granada, high- and low energy wave heights were considered (3.1 and 0.5 m, respectively). Two directions were selected to represent the bidirectional wave climate of Playa Granada (West, 238°, and East, 107°). These combination of wave heights and directions have been used as representative conditions of the wave climate in Playa Granada un different research works [4, 29, 30, 32]. In order to simulate storm conditions, the duration of this sea state for the assessment of the shoreline position was 48 h.

The third-generation wave propagation model applied was Delft3D-Wave, which is based on SWAN [17]. The propagation was made by means of two nested grid: a coarse grid (82 × 82 cells) whose cell sizes varied from 170 × 65 m in the offshore region to 80 × 80 m in the coastline, and a finer grid (244 × 82 cells), with cell sizes about 25 × 15 m. The frequency space was divided into 37 logarithmically distributed frequencies, from 0.03 to 1 Hz and 72 directions were taken into account, covering the whole circle in increments of 5°. This numerical model have been previously used successfully in Playa Granada [12].

The shape and configuration of the wave farm were defined by its location, the rows in which the devices are installed and the spacing between them. In total, we studied 6 locations, WECs arranged in 1–4 rows and separated by 4 different distances, from 1D to 4D (D = 90 m) which, with the sea states defined in the previous paragraph, yields 384 cases. The number of devices forming each wave farm was kept constant (11). The different configurations and locations were implemented in the wave propagation model in order to study the modifications made on wave patterns by each one.

In order to simulate the wave-structure interaction, the wave farms were incorporated as obstacles in the wave propagation model by means of the reflection ($K_r$) and transmission ($K_t$) coefficients of the devices. WaveCat, an overtopping type device, has been selected for this work as its efficiency for coastal defence purposes has been widely proved [1, 8, 29, 30, 32]. This device is composed by two hulls—which can

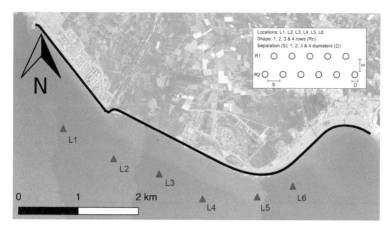

**Fig. 2.5** Wave farm locations and layouts implemented on the numerical models (*Source* [31]. Reproduced with permission of Elsevier)

be moved—joined at the stern. Reference [13] obtained transmission and reflection coefficients for this device by means of an extensive laboratory campaign.

Based on the breaking parameters obtained from the wave propagation model, [33] formulation was applied in order to obtain LST rates. This equation was considered as it has been developed to be used for all kinds of coasts including sandy, gravel and shingle beaches. This formulation is expressed as:

$$Q_m = 0.00018 K_{swell} \rho_s g^{0.5} (\tan \beta)^{0.4} (D_{50})^{-0.6} (H_{s,br})^{3.1} \sin(2\theta_{br}), \qquad (2.3)$$

with $Q_m$ the LST rate, $K_{swell}$ a coefficient which takes into account the effect of long periods waves (swell waves), $\rho_s$ the sediment density, $g$ the acceleration of gravity, $\tan \beta$ the bed slope of the surf zone, $D_{50}$ the size of the sediment, $H_{s,br}$ the significant wave height at breaking and $\theta_{br}$ the angle between the shoreline and the wave front.

The movement of the shoreline position produced by the LST was studied applying a one-line model [28]. This model make a sediment balance based on LST rates and assess the evolution of the shoreline position. Its expression is as follows:

$$\frac{\partial y_s}{\partial t} = \frac{1}{D_c} \left( -\frac{\partial Q_t}{\partial x} \right), \qquad (2.4)$$

with $y_s$ the shoreline position and $Q_t$ the LST rate. $D_c$ accounts for the active zone of the beach profile and is normally taken as the sum of the depths of closure and the height of the berm. The ability of [33] formulation coupled with a one-line model to reproduce the movement of the shoreline in Playa Granada has been validated in previous works [12]. Once the final shoreline position has been obtained for the 384 cases, the dry beach surface difference ($\Delta A$) between the baseline scenario and each case was obtained to generate the target values of the dataset.

### 2.3.3  ANN Implementation

The dataset for the training was defined using the numerical models described in the previous section. The output vector is formed by only one variable: the dry beach surface difference ($\Delta A$) as indicator of the coastal protection provided by the wave farm. The input vector was formed by (Fig. 2.5): the alongshore location ($L$), the spacing between devices ($S$), the rows composing the farm ($R_o$), the significant wave height ($H_s$) and the mean wave direction ($\theta$). Then, the ANN model may be expressed as:

$$\Delta A = f_{ANN} (L, R_o, S, H_s, \theta) . \qquad (2.5)$$

Between the cases forming the dataset, 264 cases ($\approx$70%) were selected to form the training dataset. Then, the test and validation datasets were formed by 60 cases ($\approx$15%) each one. In order to ensure that the test process account correctly for the performance of the ANN, for each one of the 6 locations studied 10 cases representative of the variety of conditions analysed were selected to be included in the test dataset. The cases for the validation dataset were selected in a similar way whereas the remaining cases were included in the training dataset.

There is no way to know the optimum architecture for an ANN before its application. The number of neurons and number, i.e. its architecture, which yield the best performance if a combination of the nature of the dataset, the number of cases and the inputs and outputs variables, which define the function we need to approximate. If a low number of neurons are used, a fitting is expected. Increasing the number of neurons the ANN performance if improved as it is able to reproduce higher order statistics of the dataset but, if we surpass a certain threshold the generalization capabilities of the network decrease and the function may not fit data outside the original training dataset [34]. This is process is known as over-fitting. Because of that, the optimum architecture for an ANN is found by means of a trial-and-error process [15].

In every ANN model, the input and output layer sizes are fixed by the number of input and output variables. So, in this case, the ANN we are looking for has 5 and 1 neurons in the input and output layer, respectively. Then, in order to find the optimum number of neurons in the hidden layer, a comparative study was carried out. In this study, one and two hidden layers architectures were analysed. On the one hand, [5-2-1], [5-5-1], [5-10-1], [5-15-1], [5-20-1], [5-25-1], [5-30-1] and [5-35-1] architectures were tested. On the other hand, 64 architectures with two hidden layers were studied. These ANNs were formed by a first layer of 5, 10, 20 and 30 neurons; and a second layer with a number of neurons ranging from 2 to 35.

The two training algorithms described in Sect. 2.3.1 (Levenberg–Marquardt and the Bayesian regularization) were incorporated to the comparative study. When an ANN is trained, the initial weights and biases are generated randomly. Then, each time an ANN is trained slightly different results are obtained. In order to avoid this effect, each architecture was trained 101 times and the averaged performance to generalize the target values of the dataset was computed. This performance was assessed by means of root-mean-squared error (RMSE), calculated as:

$$\text{RMSE} = \sqrt{\frac{\sum_{i=1}^{N}(e_i - m_i)^2}{N}}, \tag{2.6}$$

with $e$ the target value, $m_i$ the output produced by the ANN and $N$ the number of cases.

### 2.3.4 Wave Farm Optimization in the Study Zone

In order to show the ability of the proposed methodology to optimize the location and layout of a wave farm for coastal defence against erosion, the ANN which yielded the best results in the comparative study was applied to the study site. The different spaces between devices and number of rows, along with the different wave heights and directions were evaluated for 100 locations between L1 and L6. The results provided by the ANN were analysed, and finally, the optimum location and configuration of the wave farm for coastal protection were selected.

## 2.4 Results

### 2.4.1 Comparison of the Different Architectures and Algorithms

In this section, the results yielded by the comparative study of different ANN architectures are presented. The average RMSE obtained by the different architectures with one hidden layer is shown in Fig. 2.6.

Using the Levenberg–Marquardt algorithm, average RMSEs for the training dataset range from $8.78\,\text{m}^2$ in the ANN with the smallest hidden layer [5-2-1] to $3.91\,\text{m}^2$ in the biggest [5-35-1]. However, the validation and test datasets show the opposite behaviour. The average RMSE is reduced from $8.69\,\text{m}^2$ ($9\,\text{m}^2$) to $5.84\,\text{m}^2$ ($5.47\,\text{m}^2$) for architecture [5-20-1] in the validation (test) dataset. From this architecture to [5-35-1] over-fitting appear, rising average RMSEs up to $6.48\,\text{m}^2$ ($6.09\,\text{m}^2$).

With the Bayesian algorithm over-fitting is even more appreciable. Average RMSE is decreased for the training dataset from $9.64\,\text{m}^2$ in [5-2-1] to $3.92\,\text{m}^2$ in [5-20-1] and, from there, the average RMSE grows up to $12.87\,\text{m}^2$ with [5-35-1]. A similar behaviour is observed also in the test dataset, which evaluate the ability of the ANN to generalise the data out of the training dataset. In this case average error decrease up to $4.77\,\text{m}^2$ for [5-10-1] architecture. From then, average RMSE grows up to a maximum of $12\,\text{m}^2$ for the architecture with the largest hidden layer [5-35-1].

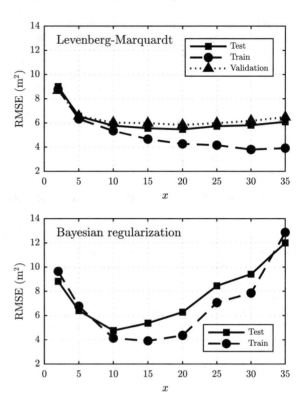

**Fig. 2.6** Average RMSE of the architectures with one hidden layer. The abscissa shows the value of $x$ in [5-$x$-1] architecture (*Source* [31]. Reproduced with permission of Elsevier)

Architectures with two hidden layers trained by means of the Levenberg–Marquardt algorithm depict also the same over-fitting problems (Fig. 2.7). The average error found in the train dataset is lower when the number of neurons in the second hidden layer is higher, with a minimum average error of $2.31 \, \text{m}^2$ found in the [5-30-35-1] architecture. Average errors are even higher for the validation and test datasets. [5-5-20-1] architecture depicts the minimum error with the test data ($5.72 \, \text{m}^2$), whereas average RMSE ranges between 7.8 and $11.6 \, \text{m}^2$ for those architectures with 30 neurons in the first hidden layer.

ANN architectures trained by means of the Bayesian regression algorithm show an even more significant over-fitting. Minimum average error in the training dataset is found with [5-20-10-1] architecture ($2.5 \cdot 10^{-5} \, \text{m}^2$), whereas average RMSE in the architectures with 30 neurons in the first hidden layer reaches the same magnitude. However, the average errors are higher for the test dataset which means a poor generalization of the data. In this case, the [5-30-20-1] architecture show the minimum average error ($5.1 \, \text{m}^2$), whereas the [5-5-35-1] architecture reaches the higher average RMSE ($8.2 \, \text{m}^2$).

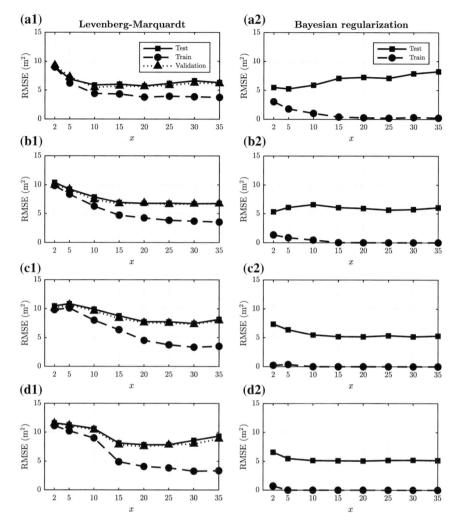

**Fig. 2.7** Average RMSE of the architectures with two hidden layers. The abscissa shows the value of $x$ in: **a** [5-5-$x$-1] architecture, **b** [5-10-$x$-1] architecture, **c** [5-20-$x$-1] architecture and **d** [5-30-$x$-1] architecture (*Source* [31]. Reproduced with permission of Elsevier)

As a result of the comparative study, a better performance of the one hidden layer architectures over the two-hidden layer ANNs is observed. The minimum average observed for the one hidden layer architectures is $4.77\,m^2$ ($5.47\,m^2$) for the ANNs trained with the Bayesian regularization (Levenberg–Marquardt) algorithm. However, minimum error increase up to $5.1\,m^2$ ($5.72\,m^2$) for the two-hidden-layer architectures. In both cases, one- and two-hidden-layer systems, Bayesian regularization algorithm shows a better performance, yielding lower RMSEs. However, the efficiency of this algorithm is lower as it requires a bigger computational effort as illustrated by the mean running time which was a 33% lower for the Levenberg–Marquardt algorithm. This greater computational performance is explained as the Levenberg–Marquardt algorithm is capable to reach the equilibrium RMSE before due to the validation process.

### 2.4.2   Final ANN Model

The results of the comparative study shown in the previous sections yielded that [5-10-1] architecture, trained by means of the Bayesian regularization algorithm, achieved the best-fit to the data. The ANN with an error closest to the median value was chosen between the 101 runs carried out during the comparative study. The error in the training process was $3.23\,\text{m}^2$. This low error is the first step for a proper performance of the neural network. The good fit provided by the chosen ANN to the training dataset is depicted in Fig. 2.8. But, more importantly, the generalization capabilities of the ANN are ensured by the low error achieved in the test dataset $(4.22\,\text{m}^2)$. This RMSE indicates that the model is capable of assess the beach surface differences even for cases out of the training dataset (Fig. 2.8b).

**Fig. 2.8**   ANN outputs versus modelled data (*Source* [31]. Reproduced with permission of Elsevier)

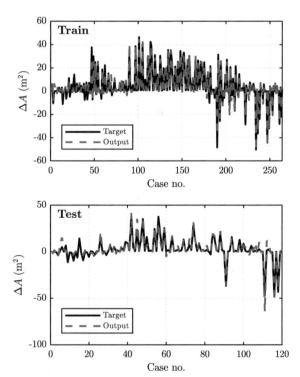

**Fig. 2.9** Linear regression for the training and test datasets (*Source* [31]. Reproduced with permission of Elsevier)

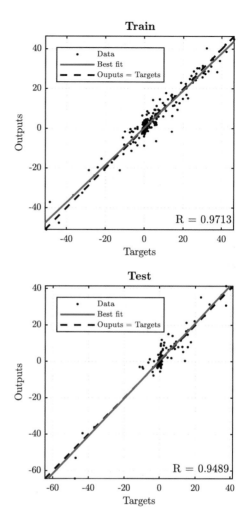

The best linear fit for the test dataset confirm these good performance of the model (Fig. 2.9). The obtained equation is actually very close to the diagonal $y = x$:

$$y_{train} = 0.9391x + 0.2705, \tag{2.7}$$

with a correlation coefficient of $R = 0.9713$. This fact proves that the model has been able to learn properly the characteristics of the training dataset.

In the case of the test dataset, the correlation coefficient shows also an excellent fit to the data ($R = 0.9489$). The yielded best linear fit is also close to the diagonal:

$$y_{test} = 1.033x + 0.3225. \tag{2.8}$$

The excellent results of the linear fit along with the low error obtained indicates the ANN model is able to predict dry beach area difference with good accuracy, even for cases with are totally new to the model. This is also ensured by the composition of the train dataset which was designed to encompass the different locations, layouts and characteristics of the wave climate studied. Based on this results, the ANN is validated and, in the next section, an optimization of the location and configuration of a wave farm in Playa Granada will be depicted.

### 2.4.3  Wave Farm Optimization

Once the best ANN architecture and training algorithm have been selected, the model was used to study 100 different locations of the modelled wave farm between L1 and L6. The layouts (number of rows and spacing between devices) and characteristics of the wave climate (significant wave heights and mean directions) were the same than those used for the training dataset.

Under westerly waves, the optimum location for the wave farm is found near the central stretch of Playa Granada (Fig. 2.10). The layout which yielded the best result $(27.4\,m^2)$ was formed by 2 rows with devices separated 180 m (2D). Wave farms situated westwards to the optimum farm yielded worse results showing a constant erosion of about $5\,m^2$ for those farms located close to Salobreña Rock. Under easterly waves, the wave farm with the better performance is located very close to that found for westerly waves but situated eastward, closer to Punta del Santo. The dry beach area difference obtained was $48.17\,m^2$ and the best layout was the same. However, those wave farms situated eastwards to the optimum one provided negative results, inducing losses of dry beach area near $60\,m^2$ as the wave farms situated in this area block the sediment transport from Poniente Beach which tend to recover the dry beach surface of Playa Granada. The farms situated to the west also yielded a positive effect over Playa Granada although the dry beach area obtained in theses cases is lower.

The dry beach area differences calculated for each configuration and alongshore location were averaged taking into account the number of westerly and easterly waves occurred during the last 25 years, as an indicator of the overall performance of the wave farm. Negative results were found for those farms situated at the east part of the study region, with a maximum erosion of $30\,m^2$ found for layouts formed by 3 rows and a spacing of 90 m $(1D)$, due to the block of the sediment transport from Poniente Beach under easterly sea states.

In the case of the wave farms situated to west end of the study site, there is almost no impact produced with dry beach surface difference near zero. The alongshore location and layout of the optimum overall wave farm is depicted in Fig. 2.10. The same layout: 2 rows and a spacing of 180 m (2D) is found to be the optimum. In this case, the best alongshore location is situated between the optimum locations found for easterly and westerly waves, slightly to the west of the best location easterly sea states.

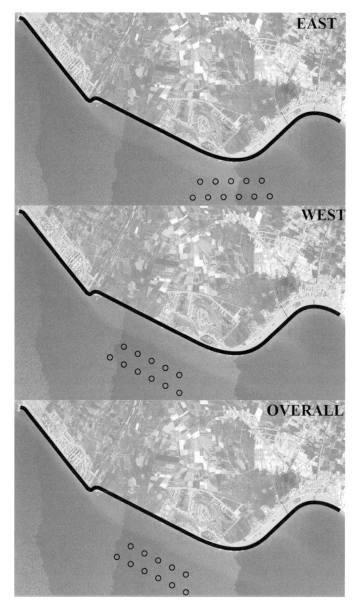

**Fig. 2.10**  Location and layout of the optimum wave farm for coastal protection for westerly/easterly waves and overall  (*Source* [31]. Reproduced with permission of Elsevier)

After the duration considered (48 h) a weighted dry beach area difference of 29.59 m$^2$ was yielded by the optimum wave farm. This means an increase in subaerial beach surface of 5400.18 m$^2$ per year with respect to the no-farm scenario. This

**Fig. 2.11** Rate of dry beach area variations: natural situation (without wave farm), protection provided by the farm and balance (with wave farm) (*Source* [31]. Reproduced with permission of Elsevier)

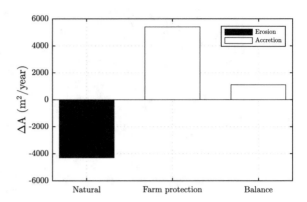

performance would help to decrease the erosion found in the study area, where near 4300 m$^2$/year have been lost since the regulation of the basin run-off [11], and would be able to turn the shoreline response to accretionary, as shown in Fig. 2.11.

The results yielded by the ANN model agree with those obtained in previous studies carried out in Playa Granada. The effects of the alongshore-position of a wave farm, with no variation in its layout, was studied by [30] and the results of this work show an optimum performance for a wave farm situated in a similar location, yielding a similar dry beach surface (25.58 m$^2$ per 48 h).

References [29, 32] studied the implications of the layout in the coastal defence performance of a wave farm in the study site and obtained very similar results to the yielded by the ANN model, with the optimum layout formed also by two rows and an inter-device spacing of 180 m (2D). This agreement between the model and the results of other research works shows the ability of the ANN to assess the coastal defence provided by wave farms.

## 2.5 Conclusions

As shown by recent research works, wave farms are able to provide coastal protection along with carbon-free energy. Many factors may affect its coastal defence performance: the layout in which the devices are arranged, the inter-device spacing, the wave climate and also the alongshore position. ANNs are artificial intelligence tools designed to provide the best fit to complicate functions learning from datasets composed by input and target vectors. In this work, one- and two-hidden-layer systems forming over forty architectures and trained using two different training algorithms were studied to benchmark its performance to predict the coastal protection provided by wave farms.

The lower errors were yielded by those architectures formed by a unique hidden layer composed by 10–20 neurons. Training and test error were higher for those architectures with a lower number of neurons in the hidden layer as they were not able to learn properly the characteristics of the dataset. Although the training error for architectures with a higher number of neurons was lower, the RMSE for the test dataset was higher which show a poor ability to generalize the results for data out of the test dataset. In a similar way, the two-hidden-layer architectures were able to yield lower errors for the train data, however the RMSE was higher for the test dataset that the obtained using one-hidden-layer architectures. Regarding the training algorithm tested, the networks trained using the Bayesian regression algorithm depicted a lower RMSE than those trained using the Levenberg–Marquardt algorithm.

The [5-10-1] architecture—trained with the Bayesian regression algorithm— provided the best results with an error of $3.23\,m^2$ ($4.22\,m^2$) for the train (test) dataset. In addition, the correlation coefficient for both train and test dataset (0.9713 and 0.9489, respectively) confirmed the best results yielded by this ANN model.

Once the optimum architecture and training algorithm for the ANN was obtained and validated, the model was used to study the best layout and location to mitigates the erosion problems in Playa Granada. The optimum location was found close to the centre of the studied beach and the layout which provided the best results in terms of coastal protection was formed by two rows of WECs separated 180 m. This results highlight the efficiency of this methodology which can be useful for stakeholders, designers and managers during a wave farm project in order to provide coastal defence along with carbon-free energy generation.

However, research in this field is still in an early stage and further work is needed in order to properly develop wave farms as coastal defence devices. Future research works could focus on aspects as the selection of a wider range of sea states to properly assess the protection provided in the long term or the influence of the composition of the beach sediments. Moreover, the mooring system or the control strategy are design variables which have not been taken into account in this work and that could be included in the ANN model in the future. In this basis, the proposed model could turn into a real-time application. In addition, the geometry of the WEC devices is an additional parameter that should be studied in the future.

**Acknowledgements** The projects, grants, funding entities and data sources that have supported this chapter are specified in the preface of the book.

# References

1. Abanades J, Flor-Blanco G, Flor G, Iglesias G (2018) Dual wave farms for energy production and coastal protection. Ocean Coast Manag 160:18–29. https://doi.org/10.1016/j.ocecoaman.2018.03.038
2. Bergillos RJ, Ortega-Sánchez M, Losada MA(2015) Foreshore evolution of a mixed sand and gravel beach: the case of Playa Granada (Southern Spain). In: Proceedings of the 8th coastal sediments. World Scientific
3. Bergillos RJ, Rodríguez-Delgado C, López-Ruiz A, Millares A, Ortega-Sánchez M, Losada MA (2015) Recent human-induced coastal changes in the Guadalfeo river deltaic system (southern Spain). In: Proceedings of the 36th IAHR-international association for hydro-environment engineering and research world congress

4. Bergillos RJ, López-Ruiz A, Ortega-Sánchez M, Masselink G, Losada MA (2016) Implications of delta retreat on wave propagation and longshore sediment transport-Guadalfeo case study (southern Spain). Mar Geol 382:1–16
5. Bergillos RJ, Masselink G, McCall RT, Ortega-Sánchez M (2016) Modelling overwash vulnerability along mixed sand-gravel coasts with XBeach-G: case study of Playa Granada, southern Spain. In: Coastal Engineering Proceedings, vol. 1 (35), p 13
6. Bergillos RJ, Ortega-Sánchez M (2017) Assessing and mitigating the landscape effects of river damming on the Guadalfeo River delta, southern Spain. Landscape Urban Plann 165:117–129
7. Bergillos RJ, Masselink G, Ortega-Sánchez M (2017) Coupling cross-shore and longshore sediment transport to model storm response along a mixed sand-gravel coast under varying wave directions. Coastal Engineering 129:93–104
8. Bergillos RJ, Lopez-Ruiz A, Medina-Lopez E, Monino A, Ortega-Sanchez M (2018) The role of wave energy converter farms on coastal protection in eroding deltas, Guadalfeo, southern Spain. J Clean Prod 171:356–367
9. Bergillos RJ, López-Ruiz A, Principal-Gómez D, Ortega-Sánchez M (2018) An integrated methodology to forecast the efficiency of nourishment strategies in eroding deltas. Sci Total Environ 613:1175–1184
10. Bergillos RJ, Ortega-Sánchez M, Masselink G, Losada MA (2016) Morpho-sedimentary dynamics of a micro-tidal mixed sand and gravel beach, Playa Granada, southern Spain. Mar Geol 379:28–38
11. Bergillos RJ, Rodríguez-Delgado C, Millares A, Ortega-Sánchez M, Losada MA (2016) Impact of river regulation on a mediterranean delta: assessment of managed versus unmanaged scenarios. Water Resour Res 52(7):5132–5148
12. Bergillos RJ, Rodríguez-Delgado C, Ortega-Sánchez M (2017) Advances in management tools for modeling artificial nourishments in mixed beaches. J Mar Syst 172:1–13
13. Fernandez H, Iglesias G, Carballo R, Castro A, Fraguela J, Taveira-Pinto F, Sanchez M (2012) The new wave energy converter WaveCat: concept and laboratory tests. Mar Struct 29:58–70
14. Hagan MT, Menhaj MB (1994) Training feedforward networks with the Marquardt algorithm. IEEE Trans Neural Netw 5(6):989–993
15. Haykin S (1994) Neural networks: a comprehensive foundation, 1st edn. Prentice Hall PTR, Upper Saddle River
16. He X, Xu S (2010) Process neural networks: Theory and applications. Springer Science & Business Media, Berlin
17. Holthuijsen L, Booij N, Ris R (1993) A spectral wave model for the coastal zone. ASCE
18. Iglesias G, Carballo R, Castro A, Fraga B (2009) Development and design of the WaveCat™ energy converter. In: Coastal engineering 2008: (in 5 volumes). World Scientific, Singapore. pp. 3970–3982
19. Jain P, Deo M (2007) Real-time wave forecasts off the western Indian coast. Appl Ocean Res 29(1–2):72–79
20. Jalón ML, Baquerizo A, Losada MA (2016) Optimization at different time scales for the design and management of an oscillating water column system. Energy 95:110–123
21. López I, Iglesias G (2014) Efficiency of OWC wave energy converters: a virtual laboratory. Appl Ocean Res 44:63–70
22. López I, Castro A, Iglesias G (2015) Hydrodynamic performance of an oscillating water column wave energy converter by means of particle imaging velocimetry. Energy 83:89–103. https://doi.org/10.1016/j.energy.2015.01.119
23. López I, López M, Iglesias G (2015) Artificial neural networks applied to port operability assessment. Ocean Eng 109:298–308
24. López I, Pereiras B, Castro F, Iglesias G (2015) Performance of OWC wave energy converters: Influence of turbine damping and tidal variability. Int J Energy Res 39(4):472–483. ER-13-4164.R2
25. López-Ruiz A, Bergillos RJ, Ortega-Sánchez M (2016) The importance of wave climate forecasting on the decision-making process for nearshore wave energy exploitation. Appl Energy 182:191–203

26. MacKay DJ (1992) Bayesian interpolation. Neural Comput 4(3):415–447
27. Ortega-Sánchez M, Bergillos RJ, López-Ruiz A, Losada MA (2017) Morphodynamics of mediterranean mixed sand and gravel coasts. Springer, Berlin
28. Pelnard-Considère, R.: Essai de theorie de l'evolution des formes de rivage en plages de sable et de galets. Les Energies de la Mer: Compte Rendu Des Quatriemes Journees de L'hydraulique, Paris 13, 14 and 15 Juin 1956; Question III, rapport 1, 74-1-10 (1956)
29. Rodriguez-Delgado C, Bergillos RJ, Ortega-Sánchez M, Iglesias G (2018) Protection of gravel-dominated coasts through wave farms: layout and shoreline evolution. Sci Total Environ 636:1541–1552
30. Rodriguez-Delgado C, Bergillos RJ, Ortega-Sánchez M, Iglesias G (2018) Wave farm effects on the coast: the alongshore position. Sci Total Environ 640:1176–1186
31. Rodriguez-Delgado C, Bergillos RJ, Iglesias G (2019) An artificial neural network model of coastal erosion mitigation through wave farms. Environ Softw Modell 119:390–399
32. Rodriguez-Delgado C, Bergillos RJ, Iglesias G (2019) Dual wave energy converter farms and coastline dynamics: the role of inter-device spacing. Sci Total Environ 646:1241–1252
33. van Rijn LC (2014) A simple general expression for longshore transport of sand, gravel and shingle. Coast Eng 90:23–39
34. Walczak S, Cerpa N (1999) Heuristic principles for the design of artificial neural networks. Inf Softw Technol 41(2):107–117

# Chapter 3
# Wave Energy Converter Configuration for Coastal Erosion Mitigation

**Abstract** This chapter analyses the influence of the wave energy converter geometry, in particular the wedge angle of WaveCat devices, on the performance of wave farms as coastal protection elements against erosion. Laboratory experiments were conducted for two angles between hulls under low-, mid- and high-energy conditions to obtain the reflection and diffraction coefficients. These values were used as input for the joint application of a wave propagation model, a longshore sediment transport formulation and the one-line model to a study site in southern Spain. The shoreline evolution and dry beach area availability for wave farms with by both devices were assessed. The results indicate that WaveCat devices with a wedge angle of 60° provide more protection than those with 30° for long wave periods and less protection for short periods. Thus, to optimize the efficiency of wave farms for coastal defence, the geometry of the wave energy converters should be adapted dynamically to the incoming wave condition.

## 3.1 Objective

This chapter is aimed at exploring the effects of the angle between the twin hulls of WaveCat devices on significant wave heights at breaking, longshore sediment transport (LST) rates, shoreline geometry and dry beach area availability. For that, laboratory testes were carried out in a wave tank (Sect. 3.3.1) to measure the transmission and reflection coefficients. These measurements were used to apply the Delft3D-Wave model (Sect. 3.3.2.1), the LST equation proposed by [21] (Sect. 3.3.2.2) and a one-line model (Sect. 3.3.2.3) to a case study in southern Iberian Peninsula (Sect. 3.2).

## 3.2 Study Area

The selected study area, Playa Granada, is a Mediterranean beach located in southern Iberian Peninsula (Fig. 3.1a). This stretch of coast is bounded to the west and to the east by the Guadalfeo River mouth and by a shoreline horn (*Punta del Santo*),

© The Author(s), under exclusive license to Springer Nature Switzerland AG 2020
R. J. Bergillos et al., *Ocean Energy and Coastal Protection*,
SpringerBriefs in Energy, https://doi.org/10.1007/978-3-030-31318-0_3

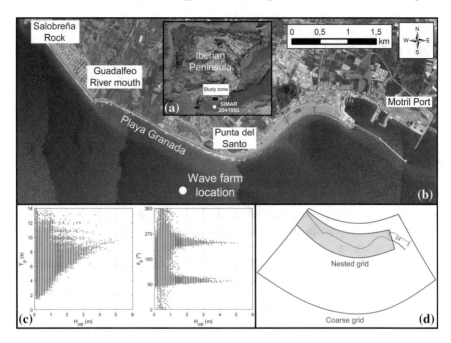

**Fig. 3.1** **a–b** Location and plan view of the study area (Playa Granada). **c** Distributions significant wave height-spectral peak period and significant wave height-mean wave direction in deep waters. **d** Boundaries of the grids defined to apply the Delft3D-Wave model (*Source* [14]. Reproduced with permission of Elsevier)

respectively (Fig. 3.1). This deltaic coast has been subjected to strong erosion issues in the past two decades [3, 7, 13], which have been aggravated since the regulation of the Guadalfeo River in 2004 [4, 6, 10].

The tidal range in the study area is ~0.6 m [16], whereas surge levels exceed 0.5 m during storms [5, 11]. The beach is subjected to two prevailing wave directions (south-west and south-east, Fig. 3.1c) and the 50%, 90% and 99.9% non-exceedance significant wave heights in deep water are 0.5 m, 1.2 m and 3.1 m, respectively [8].

## 3.3  Methodology

### 3.3.1  Laboratory Tests

The transmission ($K_t$) and reflection ($K_r$) coefficients for two wedge angles ($\alpha = 30°$ and $\alpha = 60°$, Fig. 3.2) were measured during laboratory experiments carried out in the Ocean Basin of Plymouth University. These angles between hulls were selected as representative of conditions in which wave power per linear metre of wave front

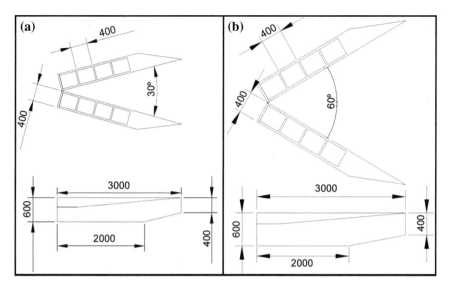

**Fig. 3.2** Dimensions (in mm) of the WaveCats tested in the laboratory: **a** $\alpha = 30°$ and **b** $\alpha = 60°$ (*Source* [14]. Reproduced with permission of Elsevier)

are substantial ($\alpha = 30°$) or limited ($\alpha = 60°$). The dimensions of the model were 3 m (length) and 0.6 m (height), representing prototype dimensions of 90 m and 18 m, respectively.

Twelve wave conditions (with a Bretschneider spectrum) were tested during the experiments: the significant wave heights and spectral peak periods ranged from 0.03 m to 0.1 m and from 1.28 s to 2.37 s, respectively. Since the experiments were conducted at a 1:30 scale and considering the Froude similarity, these values represent prototype significant wave heights from 1 to 3 m and prototype spectral peak periods from 7 to 13 s. The tested sea states are representative of the wave conditions in the study area (Fig. 3.1c). The transmission and reflection coefficients for each sea state are indicated in Table 3.1. For a detailed description of the laboratory tests, the reader is refereed to [1, 2].

### 3.3.2 Numerical Models

#### 3.3.2.1 Delft3D-Wave Model

The prototype sea states indicated in Table 3.1 were propagated from deep waters toward the nearshore region through the Delft3D-Wave model. They were modelled for the two prevailing mean wave directions at the study area (SW and SE, Fig. 3.1c). Two different grids were defined in the Delft3D-Wave model: a coarse grid covering, that extends from deep water to shallow the coastline, and a nested grid, with higher

**Table 3.1** Reflection ($K_r$) and transmission ($K_t$) coefficients for the selected angles between hulls ($\alpha$) and the wave conditions tested in the laboratory ($H_{m0,\text{mod}}$, $T_{p,\text{mod}}$). The prototype values of wave variables ($H_{m0,\text{prot}}$, $T_{p,\text{prot}}$) are also indicated

| Test case | $H_{m0,\text{mod}}$ | $H_{m0,\text{prot}}$ (m) | $T_{p,\text{mod}}$ (s) | $T_{p,\text{prot}}$ (s) | $\alpha$ (°) | $K_r$ (-) | $K_t$ (-) |
|---|---|---|---|---|---|---|---|
| S1_30 | 0.03 | 1 | 1.28 | 7 | 30 | 0.558 | 0.271 |
| S2_30 | 0.03 | 1 | 1.64 | 9 | 30 | 0.436 | 0.368 |
| S3_30 | 0.03 | 1 | 2.01 | 11 | 30 | 0.329 | 0.413 |
| S4_30 | 0.03 | 1 | 2.37 | 13 | 30 | 0.268 | 0.441 |
| S5_30 | 0.07 | 2 | 1.28 | 7 | 30 | 0.49 | 0.293 |
| S6_30 | 0.07 | 2 | 1.64 | 9 | 30 | 0.399 | 0.363 |
| S7_30 | 0.07 | 2 | 2.01 | 11 | 30 | 0.326 | 0.414 |
| S8_30 | 0.07 | 2 | 2.37 | 13 | 30 | 0.266 | 0.439 |
| S9_30 | 0.1 | 3 | 1.28 | 7 | 30 | 0.428 | 0.304 |
| S10_30 | 0.1 | 3 | 1.64 | 9 | 30 | 0.361 | 0.359 |
| S11_30 | 0.1 | 3 | 2.01 | 11 | 30 | 0.322 | 0.415 |
| S12_30 | 0.1 | 3 | 2.37 | 13 | 30 | 0.265 | 0.437 |
| S1_60 | 0.03 | 1 | 1.28 | 7 | 60 | 0.726 | 0.28 |
| S2_60 | 0.03 | 1 | 1.64 | 9 | 60 | 0.499 | 0.359 |
| S3_60 | 0.03 | 1 | 2.01 | 11 | 60 | 0.277 | 0.381 |
| S4_60 | 0.03 | 1 | 2.37 | 13 | 60 | 0.213 | 0.387 |
| S5_60 | 0.07 | 2 | 1.28 | 7 | 60 | 0.627 | 0.274 |
| S6_60 | 0.07 | 2 | 1.64 | 9 | 60 | 0.351 | 0.342 |
| S7_60 | 0.07 | 2 | 2.01 | 11 | 60 | 0.254 | 0.382 |
| S8_60 | 0.07 | 2 | 2.37 | 13 | 60 | 0.186 | 0.399 |
| S9_60 | 0.1 | 3 | 1.28 | 7 | 60 | 0.567 | 0.269 |
| S10_60 | 0.1 | 3 | 1.64 | 9 | 60 | 0.399 | 0.336 |
| S11_60 | 0.1 | 3 | 2.01 | 11 | 60 | 0.262 | 0.375 |
| S12_60 | 0.1 | 3 | 2.37 | 13 | 60 | 0.189 | 0.396 |

resolution, which covers the nearshore region (Fig. 3.1d). The Delft3D-Wave model was calibrated for the study area by means of comparison with field data collected during more than 40 days [9].

The position of the wave farm was chosen on the basis of previous studies, which proved that this location is optimum for both wave energy generation [15] and coastal defence [12, 19]. The selected layout consisted of 11 WaveCats distributed in two rows and with a inter-device spacing equal to 180 m. This geometrical configuration was also chosen based on previous works at the study area [18, 20].

The devices were introduced in the Delft3D-model as artificial obstacles. The transmission and reflection coefficients for each sea state and wedge angle were defined based on the laboratory measurements (Table 3.1). The results obtained with Delft3D-Wave were used to analyse the significant wave heights at breaking conditions and as to properly apply the LST equation of [21], which is further detailed in the following section.

### 3.3.2.2 Longshore Sediment Transport Formulation

The formulation proposed by [21] was used to obtain LST rates. This formulation can be expressed as follows:

$$Q = 0.00018 K \rho_s g^{0.5} (\tan \beta)^{0.4} (d_{50})^{-0.6} (H_{m,br})^{3.1} \sin (2\theta_{br}), \qquad (3.1)$$

where $Q$ is the LST rate, $H_{m,br}$ is the significant wave height at breaking, $\theta_{br}$ is the wave angle from shore-normal at breaking respect, $\tan \beta$ is the beach slope at breaking, $K$ is a coefficient that considers the effect of wave period on LST, $d_{50}$ is the grain size, $\rho_s$ is the sediment density and $g$ is the acceleration of gravity.

### 3.3.2.3 One-Line Model

The variations in the shoreline geometry and dry beach area availability were obtained through the LST rates computed with the equation of [21] and the application of a one-line model [17]. The latter is based on the equation:

$$\frac{\partial y_s}{\partial t} = -\frac{1}{D} \left( \frac{\partial Q}{\partial x_s} \right), \qquad (3.2)$$

where $Q$ is the LST rate, $t$ is the time, $x_s$ and $y_s$ are the shoreline coordinates, and $D$ is the sum of the closure depth height of the berm. It was demonstrated that the application of the Delft3D-Wave model (Sect. 3.3.2.1), the LST equation proposed by [21] (Sect. 3.3.2.2) and the one-line model reproduces the morphological changes of the shoreline in the study area [9].

## 3.4 Results

### 3.4.1 Significant Wave Heights at Breaking

In this section, the wave farm effects on breaking significant wave heights for the two wedge angles ($\alpha = 30°$ and $\alpha = 60°$) are analysed. Figure 3.3 depicts the differences in significant wave heights at breaking, along the studied stretch of beach, induced by wave farms with $\alpha = 30°$ and $\alpha = 60°$.

Under incoming south-westerly waves, in general, the differences are positive for long wave periods ($T_p = 11$ s and $T_p = 13$ s) and negative for short periods ($T_p = 7$ s). The maximum differences are located at the eastern part of Playa Granada (Fig. 3.3). This is induced by wave farm location (Fig. 3.1) and its influence in the wave breaking patterns in the lee of the farm.

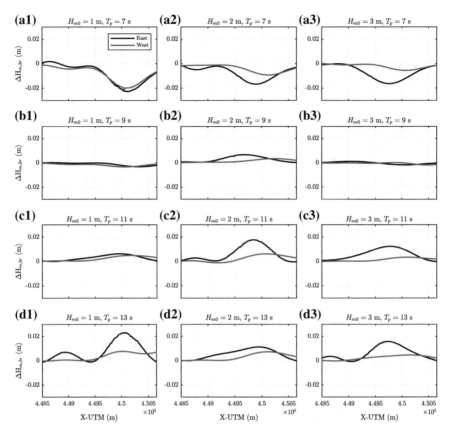

**Fig. 3.3** Variations along the studied stretch of beach of the differences in breaking significant wave heights between the wave farms with $\alpha = 30°$ and $\alpha = 60°$ under south-easterly (black) and south-westerly (red) waves. [$\Delta H_{m,br} = H_{m,br,30} - H_{m,br,60}$] (*Source* [14]. Reproduced with permission of Elsevier)

As shown in Fig. 3.4, under south-westerly waves and for constant spectral peak periods, the alongshore-averaged values of the differences in significant wave heights at breaking between $\alpha = 30°$ and $\alpha = 60°$ decrease with increasing values of the deep-water significant wave height. On the other hand, for constant values of deep-water significant wave heights, the alongshore-averaged differences under south-westerly waves increase as the spectral peak periods increases.

Therefore, the wave farm with devices with a wedge angle of 60° provides less (more) protection, in terms of wave energy at the breaking zone, for long (short) spectral peak periods than the wave farm composed by devices with $\alpha = 30°$. These differences are influenced by the different transmission and reflection coefficients of the devices with the two wedge angles (Table 3.1).

Under south-easterly wave conditions, the differences in significant wave heights are again negative for short periods and positive for long periods. As may be observed

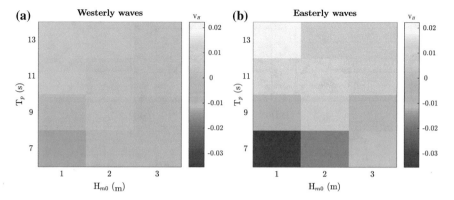

**Fig. 3.4** Variations in the alongshore-averaged values of breaking significant wave heights for the wave farm with $\alpha = 30°$ with respect to the values for $\alpha = 60°$: **a** south-westerly waves, **b** south-easterly waves. $[V_H = (\bar{H}_{m,br,30} - \bar{H}_{m,br,60})/\bar{H}_{m,br,30}]$ (*Source* [14]. Reproduced with permission of Elsevier)

in Fig. 3.3, in this case, these differences extend along most of the coastline section. For constant values of significant wave height in deep water, the alongshore-averaged differences are higher for greater values of the spectral peak period (Fig. 3.4). The differences reported in this paragraph are also induced by the different transmission and reflection coefficients for the devices with $\alpha = 30°$ and $\alpha = 60°$ (Table 3.1).

Hence, the protection in terms of wave energy at breaking provided by WaveCats with $\alpha = 60°$ respect to devices with $\alpha = 30°$ increases for increasing values of the spectral peak period under both south-westerly and south-easterly wave conditions. The differences in significant wave height at breaking between both devices under south-easterly waves are generally greater than those under south-westerly wave conditions (Fig. 3.4).

### 3.4.2 Longshore Sediment Transport Rates

This section details and compares the LST rates obtained with wave farms composed by devices with $\alpha = 30°$ and $\alpha = 60°$. The alongshore differences in LST rates between both wedge angles for the wave conditions tested in the laboratory are shown in Fig. 3.5.

For incoming south-westerly wave conditions, the differences are generally higher for decreasing spectral peak periods and increasing significant wave heights. Thus, these differences increase as the wave steepness increases. As may be observed in Fig. 3.5, the differences are more significant in the eastern part of the coast for long peak periods and in the western part for the shortest peak period.

Under SE waves, the differences generally increase with increasing values of the deep-water significant wave height (Table 3.2). The differences are more significant

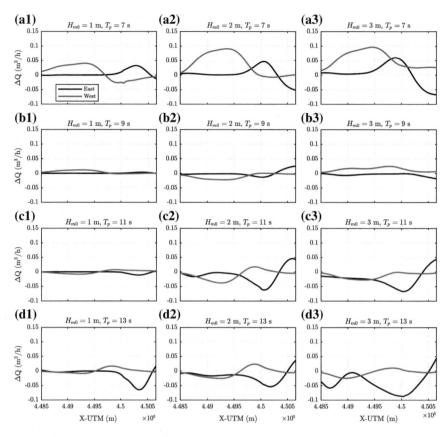

**Fig. 3.5** Variations along the studied strecth of beach of the differences in LST rates between the wave farms with $\alpha = 30°$ and $\alpha = 60°$ under south-easterly (black) and south-westerly (red) waves. $[\Delta Q = Q_{30} - Q_{60}]$ (*Source* [14]. Reproduced with permission of Elsevier)

**Table 3.2** Differences in the alongshore-averaged values of LST rates between wave farms with $\alpha = 30°$ and $\alpha = 60°$ under south-westerly and south-easterly wave conditions (in m³/h)

|  | SW waves | | | SE waves | | |
|---|---|---|---|---|---|---|
|  | $H_{m0} = 1$ m | $H_{m0} = 2$ m | $H_{m0} = 3$ m | $H_{m0} = 1$ m | $H_{m0} = 2$ m | $H_{m0} = 3$ m |
| $T_p = 7$ s | −0.0092 | −0.0385 | −0.0548 | −0.0055 | −0.0052 | −0.0073 |
| $T_p = 9$ s | −0.0045 | 0.0105 | −0.0136 | −0.0006 | 0.001 | 0.0044 |
| $T_p = 11$ s | 0.0006 | 0.0104 | 0.0134 | 0.002 | 0.0128 | 0.0256 |
| $T_p = 13$ s | −0.0007 | 0.0054 | 0.0062 | 0.0116 | 0.0185 | 0.0425 |

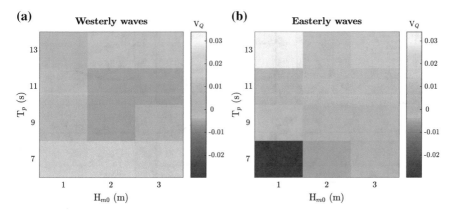

**Fig. 3.6** Variations in the alongshore-averaged values of LST rates for the wave farm with $\alpha = 30°$ with respect to the values for $\alpha = 60°$: **a** south-westerly waves, **b** south-easterly waves. [$V_Q = (\bar{Q}_{30} - \bar{Q}_{60})/\bar{Q}_{30}$] (*Source* [14]. Reproduced with permission of Elsevier)

in the lee of the farm, that is, along the eastern part of the studied stretch of coast (Fig. 3.5).

Overall, as shown in Table 3.2 and Fig. 3.6, the differences in LST rates between the wave farms composed by WaveCats with wedge angles of $\alpha = 30°$ and $\alpha = 60°$ under south-easterly waves are higher than under south-westerly wave conditions. This fact is induced by the greater differences in significant wave heights at breaking (Sect. 3.4.1) and the higher angles from shore-normal under south-easterly waves, which result in greater LST rates and differences.

### 3.4.3 Shoreline Geometry

In this section, the morphological changes induced by the wave conditions shown in Table 3.1 over a period of a month for $\alpha = 30°$ and $\alpha = 60°$ are analysed. These changes were obtained based on the LST rates reported in the previous section. Figure 3.7 depicts the differences in final shoreline geometry for the two angles between hulls under south-westerly and south-easterly wave conditions.

Under south-westerly wave conditions, the differences along the western part of Playa Granada for short wave periods are positive. This indicates that the wave farm composed by WaveCats angles between hulls of $\alpha = 60°$ induces higher accretion changes at the vicinity of the river mouth. This stretch of beach has been particularly affected by erosion issues in recent years.

The maximum differences for such short peak periods are negative, focusing on central part of the beach (Fig. 3.7). Thus, the wave farm composed by devices with a wedge angle of $\alpha = 30°$ provides more protection against shoreline erosion at this coastal section, which allocates the main urban occupations.

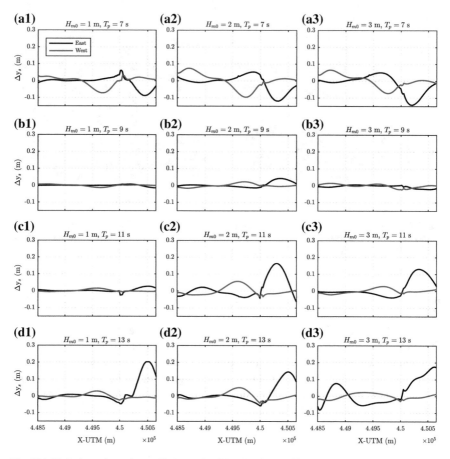

**Fig. 3.7** Variations along the studied strecth of beach of the differences between the final coastline positions (after 1 month) for the wave farms with $\alpha = 30°$ and $\alpha = 60°$ under south-easterly (black) and south-westerly (red) waves. [$\Delta y = y_{\text{final},30} - y_{\text{final},60}$] (*Source* [14]. Reproduced with permission of Elsevier)

On the contrary, the maximum differences for long peak periods under south-westerly waves are also reaches at the central part of the beach, but they are positive for these wave periods ($T_p = 11$ s and $T_p = 13$ s). As the values for long periods in Table 3.3 indicate, the farm composed by devices with $\alpha = 60°$ provides more protection against coastal erosion under these wave conditions (i.e., long periods and south-westerly wave direction).

Under south-easterly wave conditions, the highest differences are reached in the lee of the wave farm, i.e., along the eastern part of Playa Granada. In general, the differences are positive for long peak periods and negative for short periods. Thus, the devices with $\alpha = 60°$ provide more protection against shoreline erosion for long wave periods, and less protection for short periods (Fig. 3.5 and Table 3.3).

**Table 3.3** Differences in the alongshore-averaged values of final coastline positions between wave farms with $\alpha = 30°$ and $\alpha = 60°$ under south-westerly and south-easterly wave conditions (in cm)

|  | SW waves | | | SE waves | | |
|---|---|---|---|---|---|---|
|  | $H_{m0} = 1$ m | $H_{m0} = 2$ m | $H_{m0} = 3$ m | $H_{m0} = 1$ m | $H_{m0} = 2$ m | $H_{m0} = 3$ m |
| $T_p = 7$ s | −0.36 | −0.06 | 0.21 | −0.66 | −1.06 | −1.65 |
| $T_p = 9$ s | −0.04 | −0.04 | 0.02 | 0.01 | 0.54 | 0.39 |
| $T_p = 11$ s | 0.05 | 0.05 | 0.01 | 0.3 | 1.65 | 1.48 |
| $T_p = 13$ s | 0.1 | 0.06 | 0.03 | 2.1 | 1.23 | 2.01 |

The differences between device geometries under south-westerly waves are lower than under south-easterly wave conditions. These differences in shoreline geometry, which are influenced by the differences in LST rates reported in Sect. 3.4.3, in turn, dictate the availability of dry beach area, as detailed in Sect. 3.4.4.

### 3.4.4  Dry Beach Area

This section details the differences between in dry beach area between the wave farms composed by devices with $\alpha = 30°$ and $\alpha = 60°$. As shown in Fig. 3.8, these differences are positive for all the wave conditions analysed. These positive differences represent beach accretion, confirming the efficiency of wave farms for coastal protection purposes.

Previous works have proven that, in the present situation (without wave farm), south-westerly wave conditions erode the beach and south-easterly waves contribute to recover it. Thus, the wave farms change the shoreline response of Playa Granada, from erosion to deposition, under south-westerly waves, and increase the accretion along this stretch of beach under south-easterly wave conditions. For both wave directions, the dry beach area variations are generally higher as the values of significant wave height and spectral peak period increase (Fig. 3.8).

Under south-westerly waves, as shown in Table 3.4 and Fig. 3.9, the wave farm composed by WaveCats with wedge angles of $\alpha = 60°$ is generally more efficient for coastal defence, with only four exceptions (S1, S2, S5 and S6, Table 3.1). Under south-easterly wave conditions, the devices with $\alpha = 60°$ provides more protection (i.e., higher accretion values), except for the shortest peak period (Table 3.4).

These results highlight that, for an optimum efficiency for coastal defence, the device wedge angle should be adjusted dynamically depending on the wave condition. If a fixed wedge angle has to be adopted, then $\alpha = 60°$ would represent a better option than $\alpha = 30°$ in this site.

In other coastal areas, this configuration should be selected depending on the local wave climate to optimize the performance of the wave farm under the prevailing wave conditions. The methodology presented in this chapter is useful and extensible to other coasts and device geometries in order to optimize the design of wave farms in terms of coastal protection.

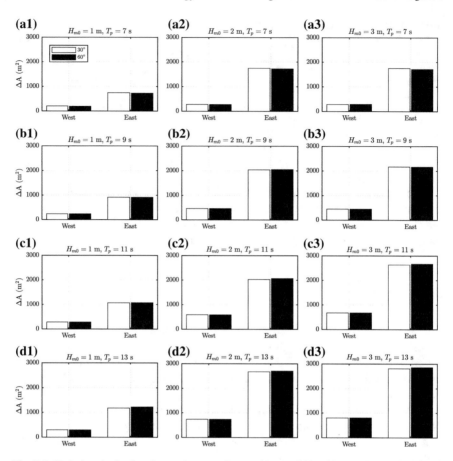

**Fig. 3.8** Variations in dry beach area for wave farms with $\alpha = 30°$ (white) and $\alpha = 60°$ (black) under south-westerly and south-easterly waves. [$\Delta A = A_{\text{final}} - A_{\text{initial}}$] (*Source* [14]. Reproduced with permission of Elsevier)

**Table 3.4** Differences in final dry beach area between wave farms with $\alpha = 30°$ and $\alpha = 60°$ under south-westerly and south-easterly waves (in m$^2$)

|  | SW waves | | | SE waves | | |
|---|---|---|---|---|---|---|
|  | $H_{m0} = 1$ m | $H_{m0} = 2$ m | $H_{m0} = 3$ m | $H_{m0} = 1$ m | $H_{m0} = 2$ m | $H_{m0} = 3$ m |
| $T_p = 7$ s | 9.4 | 1.8 | −4.7 | 15.4 | 26 | 39 |
| $T_p = 9$ s | 1 | 0.9 | −0.5 | −0.5 | −13 | −9 |
| $T_p = 11$ s | −1.3 | −1.1 | −0.1 | −7 | −41 | −36 |
| $T_p = 13$ s | −2.3 | −1.4 | −0.7 | −49 | −29 | −47 |

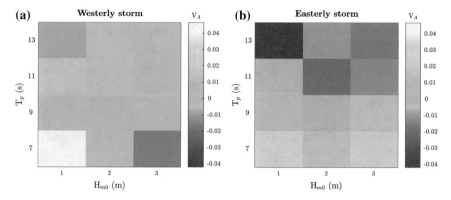

**Fig. 3.9** Variations in the dry beach area differences for the wave farm with $\alpha = 30°$ with respect to the values for $\alpha = 60°$: **a** south-westerly waves, **b** south-easterly waves. $[V_A = (\Delta A_{30} - \Delta A_{60})/\Delta A_{30}]$ (*Source* [14]. Reproduced with permission of Elsevier)

## 3.5  Conclusions

In this chapter, the effects of the wedge angle of WaveCat devices on the efficiency of wave farms for coastal protection are analysed. Two angles were considered ($\alpha = 30°$ and $\alpha = 60°$) and their influence on breaking significant wave height, LST rates, coastline morphology and dry beach area was quantified.

For that, laboratory experiments were carried out in a wave tank to determine the reflection and transmission coefficients under low-, mid- and high-energy wave conditions. These coefficients were used as inputs for the Delf3D-Wave model. The results of this model allowed computing LST rates that, in turn, were used to determine the shoreline response (through the one-line model) and dry beach area.

It is demonstrated that, in terms of breaking wave heights, the wave farm composed by devices with a wedge angle of $\alpha = 60°$ provides more protection for long wave periods and less protection for the shortest wave period considered.

The differences in breaking significant wave heights under south-easterly wave conditions are higher than under south-westerly waves. The relative angle between wave direction and shoreline orientation is also greater for south-easterly waves. These two facts lead to higher differences in LST rates between both angles between hulls under south-easterly wave conditions.

In terms of dry beach area availability, the performance of wave farms composed by WaveCats with 60° is again generally lower for the shortest peak periods ($T_p = 7$ s), but higher for long peak periods ($T_p = 11$ s and $T_p = 13$ s).

Thus, the efficiency of wave farms for coastal protection depends on the wave condition and wedge angle considered. Hence, to optimize the performance of wave farms in terms of dry beach area availability, the wedge angle should be adjusted to the sea state in a dynamic manner. The methodology detailed and applied in this chapter allows quantifying and comparing the performance of different device geometries under varying wave conditions.

**Acknowledgements** The projects, grants, funding entities and data sources that have supported this chapter are specified in the preface of the book. We thank James Allen for his support with the laboratory experiments.

# References

1. Allen J, Sampanis K, Wan J, Greaves D, Miles J, Iglesias G (2016) Laboratory tests in the development of WaveCat. Sustainability 8(12):1339
2. Allen J, Sampanis K, Wan J, Miles J, Greaves D, Iglesias G (2017) Laboratory tests and numerical modelling in the development of WaveCat. In: Proceedings of the twelfth European wave and tidal energy conference. EWTEC
3. Bergillos RJ, Ortega-Sánchez M, Losada MA (2015) Foreshore evolution of a mixed sand and gravel beach: The case of Playa Granada (southern Spain). In: Proceedings of the 8th coastal sediments. World Scientific
4. Bergillos RJ, Rodríguez-Delgado C, López-Ruiz A, Millares A, Ortega-Sánchez M, Losada MA (2015) Recent human-induced coastal changes in the Guadalfeo river deltaic system (southern Spain). In: Proceedings of the 36th IAHR-international association for hydro-environment engineering and research world congress
5. Bergillos RJ, Ortega-Sánchez M, Masselink G, Losada MA (2016) Morpho-sedimentary dynamics of a micro-tidal mixed sand and gravel beach, Playa Granada, southern Spain. Mar Geol 379:28–38
6. Bergillos RJ, Rodríguez-Delgado C, Millares A, Ortega-Sánchez M, Losada MA (2016) Impact of river regulation on a mediterranean delta: assessment of managed versus unmanaged scenarios. Water Resour Res 52(7):5132–5148
7. Bergillos RJ, López-Ruiz A, Ortega-Sánchez M, Masselink G, Losada MA (2016) Implications of delta retreat on wave propagation and longshore sediment transport-Guadalfeo case study (southern Spain). Mar Geol 382:1–16
8. Bergillos RJ, Masselink G, McCall RT, Ortega-Sánchez M (2016) Modelling overwash vulnerability along mixed sand-gravel coasts with XBeach-G: case study of Playa Granada, southern Spain. In: Coastal Engineering Proceedings, vol 1(35), p 13
9. Bergillos RJ, Rodríguez-Delgado C, Ortega-Sánchez M (2017) Advances in management tools for modeling artificial nourishments in mixed beaches. J Mar Syst 172:1–13
10. Bergillos RJ, Ortega-Sánchez M (2017) Assessing and mitigating the landscape effects of river damming on the Guadalfeo River delta, southern Spain. Landscape Urban Plann 165:117–129
11. Bergillos RJ, Masselink G, Ortega-Sánchez M (2017) Coupling cross-shore and longshore sediment transport to model storm response along a mixed sand-gravel coast under varying wave directions. Coast Eng 129:93–104
12. Bergillos RJ, Lopez-Ruiz A, Medina-Lopez E, Monino A, Ortega-Sanchez M (2018) The role of wave energy converter farms on coastal protection in eroding deltas, Guadalfeo, southern Spain. J Clean Prod 171:356–367
13. Bergillos RJ, López-Ruiz A, Principal-Gómez D, Ortega-Sánchez M (2018) An integrated methodology to forecast the efficiency of nourishment strategies in eroding deltas. Sci Total Environ 613:1175–1184
14. Bergillos RJ, Rodriguez-Delgado C, Allen J, Iglesias G (2019) Wave energy converter configuration in dual wave farms. Ocean Eng 178:204–214
15. López-Ruiz A, Bergillos RJ, Ortega-Sánchez M (2016) The importance of wave climate forecasting on the decision-making process for nearshore wave energy exploitation. Appl Energy 182:191–203
16. Ortega-Sánchez M, Bergillos RJ, López-Ruiz A, Losada MA (2017) Morphodynamics of mediterranean mixed sand and gravel coasts. Springer, Berlin

17. Pelnard-Considère R (1956) Essai de theorie de l'evolution des formes de rivage en plages de sable et de galets. Les Energies de la Mer: Compte Rendu Des Quatriemes Journees de L'hydraulique, Paris 13, 14 and 15 Juin 1956; Question III, rapport 1, 74-1-10
18. Rodriguez-Delgado C, Bergillos RJ, Ortega-Sánchez M, Iglesias G (2018) Protection of gravel-dominated coasts through wave farms: layout and shoreline evolution. Sci Total Environ 636:1541–1552
19. Rodriguez-Delgado C, Bergillos RJ, Ortega-Sánchez M, Iglesias G (2018) Wave farm effects on the coast: the alongshore position. Sci Total Environ 640:1176–1186
20. Rodriguez-Delgado C, Bergillos RJ, Iglesias G (2019) Dual wave energy converter farms and coastline dynamics: the role of inter-device spacing. Sci Total Environ 646:1241–1252
21. van Rijn LC (2014) A simple general expression for longshore transport of sand, gravel and shingle. Coast Eng 90:23–39

# Chapter 4
# Wave Energy Converter Configuration for Coastal Flooding Mitigation

**Abstract** In this chapter, the effects of wave energy converter geometry on coastal flooding are explored. In particular, it is assessed the efficiency of two angles between hulls of the WaveCat devices (30° and 60°) for the mitigation of coastal inundation. The case study consists of a wave farm composed by 11 devices deployed off a deltaic beach is southern Spain (Playa Granada). First, laboratory tests were performed to determine the reflection and transmission coefficients under low-, mid- and high-energy conditions. Then, these coefficients were used to jointly apply Delft3D-Wave and XBeach-G in Playa Granada considering wave farms with both wedge angles. The results highlight that devices with wedge angles of 60° are more efficient than those with 30° to reduce nearshore wave heights, run-up values and flooded dry beach areas for long wave periods and high-energy conditions. Thus, since coastal flooding commonly occurs under storm conditions, the devices with wedge angles of 60° are more efficient as coastal defence elements against flooding.

## 4.1 Objective

This chapter is aimed at analysing the implications of the wedge angle of WaveCat devices on wave run-up, flooded cross-shore distances and flooded area. For that, laboratory experiments were conducted in the Ocean Basin of the University of Plymouth to determine the reflection and transmission and coefficients of the WaveCat device for two wedge angles under different wave conditions (Sect. 4.3.1). These coefficients were utilized as input variables for the application of a wave propagation model (Delft3D-Wave) and coastal flooding model (XBeach-G), detailed in Sects. 4.3.2.1 and 4.3.2.2, respectively. The study area selected as case study is detailed in Sect. 4.2.

© The Author(s), under exclusive license to Springer Nature Switzerland AG 2020       45
R. J. Bergillos et al., *Ocean Energy and Coastal Protection*,
SpringerBriefs in Energy, https://doi.org/10.1007/978-3-030-31318-0_4

## 4.2  Study Area

The selected study area, known as Playa Granada, is located on the southen coast of the Iberian Peninsula, facing the Alborán Sea [5, 9]. This stretch of beach, which is the central coastal section of the deltaic coast, is bounded to the east by a shoreline horn known as *Punta del Santo* and to the west by the Guadalfeo River mouth (Fig. 4.1).

The Guadalfeo River supplies the main sediment contributions to the deltaic coast. Nevertheless, this river was dammed in 2004, regulating 85% of the basin run-off [18]. The dam represents a barrier to the natural sediment transport and, as a result, the sediment supplies from the river to the coastal area have decreased significantly reduced [8, 12, 13].

The wave climate at this coastal region is characterized by two predominant wave directions: south-east and south-west [4]. The significant wave heights in deep waters which are not exceeded 50%, 90% and 99.9% of the time are 0.5 m, 1.2 m and 3.1 m, respectively. Surge levels typically exceed 0.5 m under storm conditions [10] and the astronomical tidal range is ~0.6 m [22].

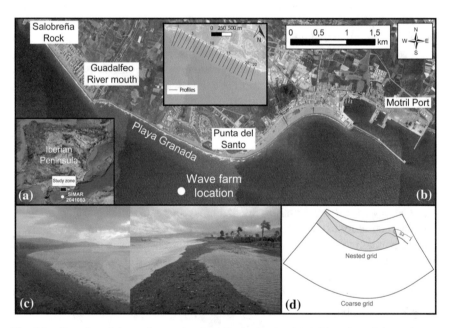

**Fig. 4.1  a** Location of the study area (southern Iberian Peninsula). **b** Plan view of the study area and location of the wave farm. **c** Images of coastal flooding in the study area. **d** Boundaries of the computational grids used in Delft3D-Wave (*Source* [11]. Reproduced with permission of Elsevier)

## 4.3 Methodology

### 4.3.1 Laboratory Tests

Laboratory tests of the WaveCat device with two angles between hulls (30° and 60°, Fig. 4.2) were conducted in the Ocean Basin of the University of Plymouth. Froude similarity and a geometrical scale of 1:30 were considered for the experiments. The model dimensions were 0.6 m high and 3 m long, representing prototype dimensions of 18 m and 90 m, respectively. A detailed description of the laboratory experiments can be consulted in [1, 2].

Twelve different sea states, representative of the wave climate at the study area, and the Bretschneider wave spectrum were selected for the experiments. The significant wave height and spectral peak period values tested in the laboratory ranged between 0.03 m and 0.1 m and between 1.28 s and 2.37 s, respectively, representing prototype values from 1 m to 3 m, and from 7 s to 13 s, respectively. The reflection and transmission coefficients were determined for all the wave conditions, as shown in Table 4.1.

### 4.3.2 Numerical Models

The Delft3D and XBeach-G models were applied to analyse the wave propagation patterns and flooded areas at the study area induced by the presence of wave farms composed by devices with wedge angles of 30° and 60°. For that, the prototype values of wave variables and the reflection-transmission coefficients shown in Table 4.1 were used (Fig. 4.1).

**Fig. 4.2** Images of the devices used for the laboratory experiments. Left: wedge angle of 30°. Right: wedge angle of 60° (*Source* [11]. Reproduced with permission of Elsevier)

**Table 4.1** Reflection coefficient ($K_r$) and transmission coefficient ($K_t$) for the sea states and the two angles between hulls tested during the laboratory experiments

|       | $H^*_{m0}$ (m) | $H_{m0}$ (m) | $T^*_p$ (s) | $T_p$ (s) | $K_{r,30°}$ (-) | $K_{t,30°}$ (-) | $K_{r,60°}$ (-) | $K_{t,60°}$ (-) |
|-------|------|------|------|------|-------|-------|-------|-------|
| SS1   | 0.03 | 1 | 1.28 | 7  | 0.558 | 0.271 | 0.726 | 0.28  |
| SS2   | 0.03 | 1 | 1.64 | 9  | 0.436 | 0.368 | 0.499 | 0.359 |
| SS3   | 0.03 | 1 | 2.01 | 11 | 0.329 | 0.413 | 0.277 | 0.381 |
| SS4   | 0.03 | 1 | 2.37 | 13 | 0.268 | 0.441 | 0.213 | 0.387 |
| SS5   | 0.07 | 2 | 1.28 | 7  | 0.49  | 0.293 | 0.627 | 0.274 |
| SS6   | 0.07 | 2 | 1.64 | 9  | 0.399 | 0.363 | 0.351 | 0.342 |
| SS7   | 0.07 | 2 | 2.01 | 11 | 0.326 | 0.414 | 0.254 | 0.382 |
| SS8   | 0.07 | 2 | 2.37 | 13 | 0.266 | 0.439 | 0.186 | 0.399 |
| SS9   | 0.01 | 3 | 1.28 | 7  | 0.428 | 0.304 | 0.567 | 0.269 |
| SS10  | 0.01 | 3 | 1.64 | 9  | 0.361 | 0.359 | 0.399 | 0.336 |
| SS11  | 0.01 | 3 | 2.01 | 11 | 0.322 | 0.415 | 0.262 | 0.375 |
| SS12  | 0.01 | 3 | 2.37 | 13 | 0.265 | 0.437 | 0.189 | 0.396 |

### 4.3.2.1  Wave Propagation Model: Delft3D-Wave

The prototype values of the sea states tested in the laboratory were propagated from deep waters toward the wave farm location with the Wave module of Delft3D, which is based on the SWAN model [16]. This model was calibrated for the study area by [14] through comparison with field measurements. The two prevailing incoming wave directions at the study area (south-east and south-west) were modelled for each sea state.

To apply the Delft3D-Wave model, two grids were defined, shown in Fig. 4.1d: a coarse computational grid, which comprise the whole deltaic region from deep water to the shoreline; and a nested grid at the nearshore region, with higher resolution to reproduce properly the interaction between wave and device. To model this interaction, the devices were defined in the model as artificial obstacles with their reflection and transmission coefficients as a function of the sea state and wedge angle considered (Table 4.1).

The selected wave farm location was proven to be optimum in terms of both wave energy availability and coastal protection by [17] and [3, 25], respectively. The wave farm layout and the inter-device spacings were also chosen on the basis of previous studies in the study area [23, 24]. The results of Delft3D-Wave at 10 m water depth were used as input wave variables to apply the XBeach-G model, detailed in Sect. 4.3.2.2.

### 4.3.2.2  Coastal Flooding Model: XBeach-G

The XBeach-G model, which was developed for gravel coasts [20, 21] and calibrated for the study site by [6, 7], was used in this work to compute wave run-up values for each wave condition shown in Table 4.1 and propagated with Delft3D-Wave for both wedge angles (Sect. 4.3.2.1).

These wave run-up values were obtained in 22 beach profiles equally distributed (one profile every 100 m) along the studied stretch of beach, as shown in Fig. 4.1. The input wave conditions for XBeach-G were the results of the Delft3D-Wave model at 10 m water depth. This offshore depth was chosen based on the recommendations proposed by the User Manual of the model [15].

The landward boundaries were variables along the studied stretch of beach depending on the occupations located land-side of each beach profile (homes, hotels, farms or golf courses). The results of XBeach-G allowed were used to compute the flooded cross-shore distances in the 22 beach profiles and total flooded area for both wedge angles under each sea state.

## 4.4 Results

### 4.4.1 Nearshore Significant Wave Heights

This sections analyse the significant wave height at 10 m water depth induced by wave farms composed by devices with the two wedge angles considered. The differences in wave heights between the case with 30° and 60° are depicted in Fig. 4.3.

For south-westerly wave conditions and short periods ($T_p = 7$ s), the differences in significant wave heights at 10 m water depth are negative (Fig. 4.3). This indicates that the wave farm composed by WaveCats with angles of 30° leads to lower nearshore wave heights under these wave conditions (Fig. 4.4). The alongshore-averaged differences in nearshore wave heights between farms composed by angles between hulls of 30° and 60° are more negative as the deep-water significant wave height increases (Table 4.2).

On the other hand, under south-westerly waves and long periods, the differences are positive along the whole stretch of beach (Fig. 4.3). This reveals that devices with an angle between hulls of 60° induce more protection in terms of nearshore wave heights. The differences increase with increasing significant wave heights and peak periods (Table 4.2 and Fig. 4.4). They are greater in the eastern part of Playa Granada (Fig. 4.3), influenced by the wave farm location (Fig. 4.1).

Under south-easterly wave conditions, the differences are again negative for the lowest spectral peak period (7 s) and positive for long periods (11 and 13 s). Thus, the wave farm composed by devices with angles between hulls of 30° induces greater nearshore wave heights (i.e. less protection) for long wave periods than the devices with 60°. The opposite occurs for the shortest peak period. The differences, which are maximum in the eastern part of the beach, decrease for decreasing values of the deep-water significant wave height.

The differences in nearshore significant wave heights are primarily induced by the different reflection and transmission coefficients for the devices with the two angles

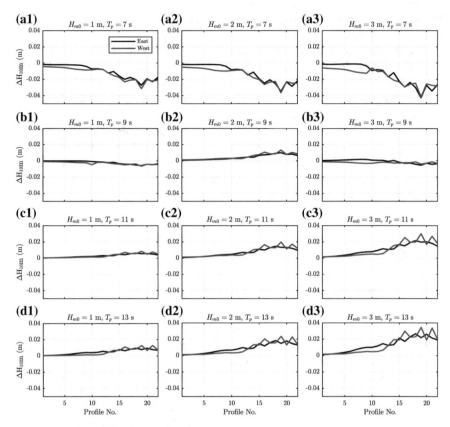

**Fig. 4.3** Variations along the studied stretch of beach of the differences in significant wave heights at 10 m water depth between the wave farms with $\alpha = 30°$ and $\alpha = 60°$ under south-easterly (black) and south-westerly (blue) waves. [$\Delta H_{10m} = H_{10m,30°} - H_{10m,60°}$] (*Source* [11]. Reproduced with permission of Elsevier)

between hulls (Table 4.1). The alongshore-averaged values of the differences under south-westerly wave conditions are slightly lower than under south-easterly waves for the shortest (less negative) and the longest (less positive) periods.

## 4.4.2  Wave Run-Up

The differences in nearshore significant wave heights reported in the previous section induce variations in wave run-up. This section analyses these variations in wave run-up between the wave farms composed by devices with both wedge angles (30° and 60°) on the 22 beach profiles shown in Fig. 4.1. Figure 4.5 depicts the alongshore

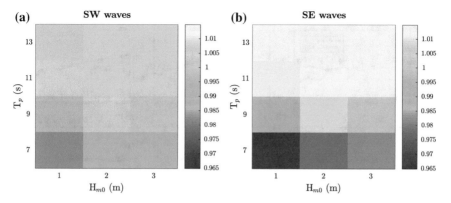

**Fig. 4.4** Variations in the alongshore-averaged values of significant wave heights at 10 m water depth for the wave farm with $\alpha = 30°$ with respect to the values for $\alpha = 60°$: **a** south-westerly waves, **b** south-easterly waves. [Variation=$\bar{H}_{10m,30°}/\bar{H}_{10m,60°}$] (*Source* [11]. Reproduced with permission of Elsevier)

**Table 4.2** Differences in the alongshore-averaged values of significant wave height at 10 m water depth between wave farms with $\alpha = 30°$ and $\alpha = 60°$ under south-westerly and south-easterly wave conditions (in cm)

| | SW waves | | | SE waves | | |
|---|---|---|---|---|---|---|
| | $H_{m0} = 1$ m | $H_{m0} = 2$ m | $H_{m0} = 3$ m | $H_{m0} = 1$ m | $H_{m0} = 2$ m | $H_{m0} = 3$ m |
| $T_p = 7$ s | −1.29 | −1.49 | −1.7 | −1.03 | −1.22 | −1.4 |
| $T_p = 9$ s | −0.3 | 0.51 | −0.23 | −0.2 | 0.43 | −0.01 |
| $T_p = 11$ s | 0.31 | 0.74 | 1.09 | 0.3 | 0.68 | 1.06 |
| $T_p = 13$ s | 0.44 | 0.85 | 1.25 | 0.48 | 0.86 | 1.28 |

variations of these differences under south-easterly and south-westerly waves for the sea states analysed.

Under south-westerly waves and for short wave periods (7 s), the differences in run-up are, in general, negative along the studied stretch of beach. This reveals that the devices with 30° are more efficient to reduce wave run-up for short wave periods.

However, as shown in Table 4.3, the differences in alongshore-averaged wave run-up values are positive for the rest of wave conditions. Thus, under mid and long wave periods, the wave farms composed by devices with wedge angles of 60° represent a better alternative for the reduction of wave run-up (Fig. 4.6).

On the other hand, for south-easterly wave conditions and short peak periods (7 s), the differences in averaged wave run-up are again negative. They are also negative for under low energy conditions for higher spectral peak periods (9 and 11 s).

Nevertheless, the WaveCats with a wedge angle of 60° are also more efficient for most of the wave conditions (Table 4.3 and Fig. 4.6), since the performance of these devices under south-easterly waves increase as the significant wave height and spectral peak period increase.

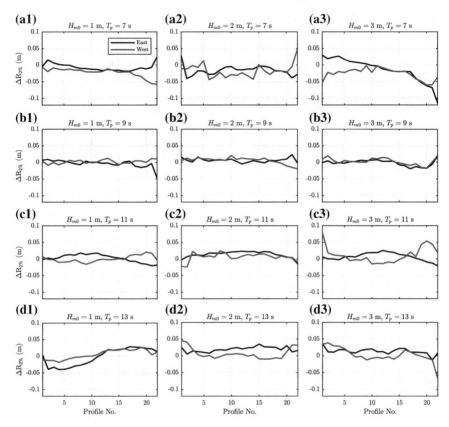

**Fig. 4.5** Variations along the studied stretch of beach of the differences in wave run-up between the wave farms with $\alpha = 30°$ and $\alpha = 60°$ under south-easterly (black) and south-westerly (blue) waves. [$\Delta R_{2\%} = R_{2\%,30°} - R_{2\%,60°}$] (*Source* [11]. Reproduced with permission of Elsevier)

**Table 4.3** Differences in the alongshore-averaged values of wave run-up between wave farms with $\alpha = 30°$ and $\alpha = 60°$ under south-westerly and south-easterly wave conditions (in cm)

|  | SW waves | | | SE waves | | |
|---|---|---|---|---|---|---|
|  | $H_{m0} = 1$ m | $H_{m0} = 2$ m | $H_{m0} = 3$ m | $H_{m0} = 1$ m | $H_{m0} = 2$ m | $H_{m0} = 3$ m |
| $T_p = 7$ s | −2.44 | −1.56 | −2.73 | −0.59 | −1.69 | −1.71 |
| $T_p = 9$ s | 0.23 | 0.48 | 0.32 | −0.54 | 0.57 | 0.14 |
| $T_p = 11$ s | 0.01 | 0.59 | 1.34 | −0.19 | 1.34 | 0.64 |
| $T_p = 13$ s | 0.58 | 0.78 | 0.01 | 0.23 | 2.01 | 1.35 |

These differences in run-up between the wave farms composed by devices with both angles induce variations in the flooded cross-shore distances and dry beach areas, as detailed in Sect. 4.4.3.

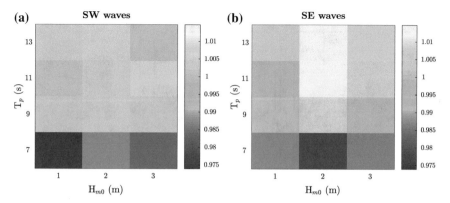

**Fig. 4.6** Variations in the alongshore-averaged values of wave run-up for the wave farm with $\alpha = 30°$ with respect to the values for $\alpha = 60°$: **a** south-westerly waves, **b** south-easterly waves. [Variation=$\bar{R}_{2\%,30°}/\bar{R}_{2\%,60°}$] (*Source* [11]. Reproduced with permission of Elsevier)

### 4.4.3 Flooded Cross-Shore Distances and Flooded Dry Beach Area

In this section, the differences between the dry beach areas which are inundated with both wedge angles are analysed. They are computed based on the wave run-up values reported in the previous section and the resulting flooded cross-shore distances in the 22 beach profiles. These flooded distances are dependent on the morphologies of the emerged profiles. Figure 4.7 represents the flooded dry beach areas under are the wave conditions analysed in this chapter and for farms composed by both wedge angles.

In all cases, the dry areas flooded under south-easterly waves are lower than under south-westerly waves, due to the shoreline orientation of the studied stretch of beach (Fig. 4.1), which is less vulnerable to incoming easterly wave conditions. Another general trend observed in Fig. 4.7 is that the higher the values of significant wave height and period, the greater the dry beach surface inundated. This trend is supported by most of the existing formulations to compute wave run-up, such as [19].

Under south-westerly waves, the differences in the areas inundated by the wave farms with both devices are negative for the shortest period (7 s), but positive for longer spectral peak periods (11 s and 13 s). Thus, the wave farms composed by WaveCats with wedge angles of 60° are more efficient to mitigate coastal flooding for long periods (Table 4.4 and Fig. 4.8). For given values of the wave period, the differences are greater (more negative or more positive) as the deep-water significant wave height increase.

Under south-easterly wave conditions, the differences between the areas inundated with farms composed by both devices are generally positive, with only three exceptions (SS1, SS2 and SS6, Tables 4.1 and 4.4). The efficiency of devices with

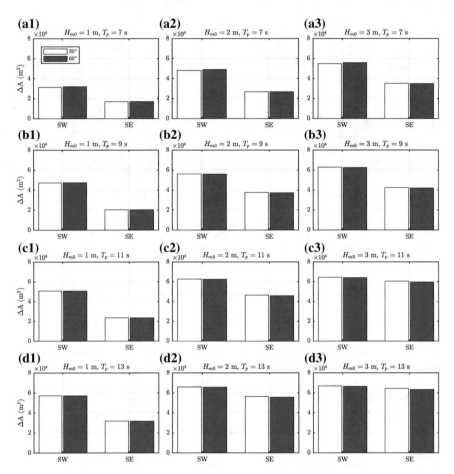

**Fig. 4.7** Total flooded areas of dry beach for the wave farms with $\alpha = 30°$ and $\alpha = 60°$ under south-easterly (SE) and south-westerly (SW) waves. (*Source* [11]. Reproduced with permission of Elsevier)

**Table 4.4** Differences in flooded dry beach area between wave farms with $\alpha = 30°$ and $\alpha = 60°$ under south-westerly and south-easterly waves (in m$^2$)

| | SW waves | | | SE waves | | |
|---|---|---|---|---|---|---|
| | $H_{m0} = 1$ m | $H_{m0} = 2$ m | $H_{m0} = 3$ m | $H_{m0} = 1$ m | $H_{m0} = 2$ m | $H_{m0} = 3$ m |
| $T_p = 7$ s | −667.05 | −953.47 | −979.39 | −91.01 | −5.86 | 299.81 |
| $T_p = 9$ s | −118.91 | 168.18 | 343.36 | −64.79 | 366.75 | 507.53 |
| $T_p = 11$ s | 36.31 | 214.06 | 415.36 | 80.82 | 519.85 | 790.48 |
| $T_p = 13$ s | 80.37 | 329.84 | 515.67 | 152.04 | 712.06 | 1018.37 |

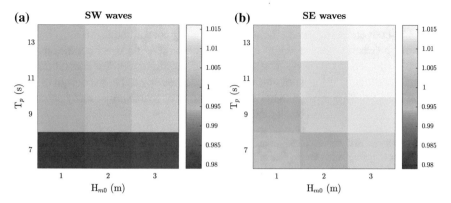

**Fig. 4.8** Variations in flooded dry beach areas for the wave farm with $\alpha = 30°$ with respect to the values for $\alpha = 60°$: **a** south-westerly waves, **b** south-easterly waves. [Variation=$A_{30°}/A_{60°}$] (*Source* [11]. Reproduced with permission of Elsevier)

wedge angles equal to 60° respect to devices with 30° is higher under south-easterly waves than under south-westerly directions for all the wave conditions.

Under storm conditions, the differences are positive under both wave directions (Table 4.4). Thus, taking into account that the higher values and problems related to coastal inundation occur for such wave conditions, the devices with wedge angles equal to 60° are clearly a more efficient strategy to protect the studied stretch of beach against coastal flooding.

## 4.5 Conclusions

This chapter deals with the influence of the device geometry on the efficiency of wave farms for coastal flooding mitigation. In particular, the effects of two wedge angles (30° and 60°) of WaveCat devices on nearshore wave height, run-up and flooded area are analysed and compared.

For that purpose, laboratory experiments were conducted in a wave tank to determine the transmission and reflections coefficients for the two angles under wave conditions representative of the case study (Playa Granada, southern Spain). The values of these coefficients were used to apply the Delft3D-Wave model and the results of this model, in turn, were utilized as input by the XBeach-G model.

It is proven that the devices with angles between hulls of 60° lead to lower values nearshore significant wave heights (i.e. greater coastal protection) for long periods and lower wave heights for short periods. These differences in wave heights, in turn, induce variations in wave run-up, so that the wedge angle of 60° is also more effective for the mitigation of run-up under long-period waves.

With respect to the beach surface inundated, the wave farms composed by devices of 60° leads to lower flooded areas under storm conditions. Therefore, taking into account that under such wave conditions flooding issues are more significant, the WaveCat converters are more efficient in alleviating coastal inundation.

Nevertheless, the wedge angle adaptation to the specific wave conditions, in a dynamic way, could allow optimizing the device configuration for coastal defence against flooding. Thus, this possibility should also be explored in the design of wave farms. This chapter offers insights that are useful for managers to mitigate coastal flooding events and their consequences.

**Acknowledgements**  The projects, grants, funding entities and data sources that have supported this chapter are specified in the preface of the book. We thank James Allen for his support with the laboratory experiments.

# References

1. Allen J, Sampanis K, Wan J, Greaves D, Miles J, Iglesias G (2016) Laboratory tests in the development of WaveCat. Sustainability 8(12):1339
2. Allen J, Sampanis K, Wan J, Miles J, Greaves D, Iglesias G (2017) Laboratory tests and numerical modelling in the development of WaveCat. In: Proceedings of the twelfth European wave and tidal energy conference. EWTEC
3. Bergillos RJ, Lopez-Ruiz A, Medina-Lopez E, Monino A, Ortega-Sanchez M (2018) The role of wave energy converter farms on coastal protection in eroding deltas, Guadalfeo, southern Spain. J Clean Prod 171:356–367
4. Bergillos RJ, López-Ruiz A, Ortega-Sánchez M, Masselink G, Losada MA (2016) Implications of delta retreat on wave propagation and longshore sediment transport-Guadalfeo case study (southern Spain). Mar Geol 382:1–16
5. Bergillos RJ, López-Ruiz A, Principal-Gómez D, Ortega-Sánchez M (2018) An integrated methodology to forecast the efficiency of nourishment strategies in eroding deltas. Scie Total Environ 613:1175–1184
6. Bergillos RJ, Masselink G, McCall RT, Ortega-Sánchez M (2016) Modelling overwash vulnerability along mixed sand-gravel coasts with XBeach-G: case study of Playa Granada, southern Spain. In: Coastal engineering proceedings, vol 1(35), p 13
7. Bergillos RJ, Masselink G, Ortega-Sánchez M (2017) Coupling cross-shore and longshore sediment transport to model storm response along a mixed sand-gravel coast under varying wave directions. Coast Eng 129:93–104
8. Bergillos RJ, Ortega-Sánchez M (2017) Assessing and mitigating the landscape effects of river damming on the Guadalfeo River delta, southern Spain. Landscape Urban Plann 165:117–129
9. Bergillos RJ, Ortega-Sánchez M, Losada MA (2015) Foreshore evolution of a mixed sand and gravel beach: the case of Playa Granada (Southern Spain). In: Proceedings of the 8th coastal sediments. World Scientific
10. Bergillos RJ, Ortega-Sánchez M, Masselink G, Losada MA (2016) Morpho-sedimentary dynamics of a micro-tidal mixed sand and gravel beach, Playa Granada, southern Spain. Mar Geol 379:28–38
11. Bergillos RJ, Rodriguez-Delgado C, Allen J, Iglesias G (2019) Wave energy converter geometry for coastal flooding mitigation. Sci Total Environ 668:1232–1241
12. Bergillos RJ, Rodríguez-Delgado C, López-Ruiz A, Millares A, Ortega-Sánchez M, Losada MA (2015) Recent human-induced coastal changes in the Guadalfeo river deltaic system (southern Spain). In: Proceedings of the 36th IAHR-international association for hydro-environment engineering and research world congress

13. Bergillos RJ, Rodríguez-Delgado C, Millares A, Ortega-Sánchez M, Losada MA (2016) Impact of river regulation on a mediterranean delta: assessment of managed versus unmanaged scenarios. Water Resour Res 52(7):5132–5148
14. Bergillos RJ, Rodríguez-Delgado C, Ortega-Sánchez M (2017) Advances in management tools for modeling artificial nourishments in mixed beaches. J Mar Syst 172:1–13
15. Deltares: XBeach-G GUI 1.0. User Manual. Delft, The Netherlands (2014)
16. Holthuijsen L, Booij N, Ris R (1993) A spectral wave model for the coastal zone. ASCE
17. López-Ruiz A, Bergillos RJ, Ortega-Sánchez M (2016) The importance of wave climate forecasting on the decision-making process for nearshore wave energy exploitation. Appl Energy 182:191–203
18. Losada MA, Baquerizo A, Ortega-Sánchez M, Ávila A (2011) Coastal evolution, sea level, and assessment of intrinsic uncertainty. J Coast Res 59:218–228
19. Matias A, Williams JJ, Masselink G, Ferreira Ó (2012) Overwash threshold for gravel barriers. Coast Eng 63:48–61
20. McCall RT, Masselink G, Poate TG, Roelvink JA, Almeida LP (2015) Modelling the morphodynamics of gravel beaches during storms with XBeach-G. Coast Eng 103:52–66
21. McCall RT, Masselink G, Poate TG, Roelvink JA, Almeida LP, Davidson M, Russell PE (2014) Modelling storm hydrodynamics on gravel beaches with XBeach-G. Coast Eng 91:231–250
22. Ortega-Sánchez M, Bergillos RJ, López-Ruiz A, Losada MA (2017) Morphodynamics of mediterranean mixed sand and gravel coasts. Springer, Berlin
23. Rodriguez-Delgado C, Bergillos RJ, Iglesias G (2019) Dual wave energy converter farms and coastline dynamics: the role of inter-device spacing. Sci Total Environ 646:1241–1252
24. Rodriguez-Delgado C, Bergillos RJ, Ortega-Sánchez M, Iglesias G (2018) Protection of gravel-dominated coasts through wave farms: layout and shoreline evolution. Sci Total Environ 636:1541–1552
25. Rodriguez-Delgado C, Bergillos RJ, Ortega-Sánchez M, Iglesias G (2018) Wave farm effects on the coast: the alongshore position. Sci Total Environ 640:1176–1186

# Chapter 5
# Management of Coastal Erosion Under Climate Change Through Wave Farms

**Abstract** In this chapter, the efficiency of wave farms in coastal protection under sea-level rise is investigated. A wave farm formed by 11 wave energy converters was modelled off Playa Granada, a gravel-dominated coast in Southern Spain, under three sea-level rise scenarios: the current water level and the water level in 2100 according to a low- and high-emission scenario. In order to explore the effects produced by the wave farm, the natural scenario without wave farm was also studied. Waves were propagated through the wave farm by means of Delft3D-Wave and breaking parameters were obtained in order to apply a longshore sediment transport (LST) formulation. The results obtained with the LST formulation were used in a one-line model to compute the changes in the position of the shoreline at the study site. The results highlight that wave farms are able to decrease beach erosion (shoreline retreat) even under sea-level rise scenarios. That makes wave farms attractive management strategies, as they contribute to the decarbonisation of the energy mix and more efficient in terms of coastal protection under sea-level rise than traditional hard-engineering structures.

## 5.1 Objective

The main objective of this chapter is to study the effects of sea-level rise on the performance of a wave farm in terms of coastal defence against shoreline retreat. To this end, three sea-level scenarios were analysed: the current, no sea-level rise, situation (baseline), and the water level in 2100 according to low-emission (RCP4.5) and high-emission (RCP8.5) projections proposed by [23]. Waves were propagated using Delft3D-Wave, a third-generation wave propagation model, for two case studies, with and without the presence of a wave farm, on Playa Granada (Southern Spain), a gravel dominated beach. An LST formulation [33] and a one-line model [27] were applied to obtain the evolution of the shoreline and the variations in dry beach area.

The structure of the chapter is as follows: Sect. 5.2 describes the study area, the methodology applied is shown in Sect. 5.3, results are presented in Sect. 5.4, a brief discussion is depicted in Sect. 5.5 and finally, main conclusions are drawn in Sect. 5.6 of this work.

R. J. Bergillos et al., *Ocean Energy and Coastal Protection*,
SpringerBriefs in Energy, https://doi.org/10.1007/978-3-030-31318-0_5

## 5.2   Study Area

Playa Granada is the central stretch of the Guadalfeo river deltaic coast. This 3-km-long beach is located on the southern coast of Iberian Peninsula that faces the Mediterranean Sea (Fig. 5.1). The Guadalfeo River mouth and its former location, *Punta del Santo*, are its west and east limits, respectively [6]. The delta is bounded to the east by Motril Port and to the west by Salobreña Rock.

The gravel fraction dominates the morphodynamic response of the beach profile and its state is reflective [9, 10]. During the last years, the stretch of beach has experienced shoreline retreat and important erosion (Fig. 5.1c), mainly due to human interventions in the river basin [5, 8]. In order to solve these problems, numerous artificial nourishment projects have been carried out over the past decade [13], however these works have not been successful in the long term [12, 14, 26].

The wave climate is clearly bimodal with two prevailing directions: west-southwest and east-southeast [11]. This is due to the passage of extra-tropical Atlantic cyclones (west direction) and Mediterranean storms (east direction). Significant wave heights during typical storms exceed 2.1 m, whereas during extreme storms 3.1 m may be achieved [7]. This coast is micro-tidal (astronomical tidal range of ∼0.6) and storm surge levels may exceed 0.5 m [10].

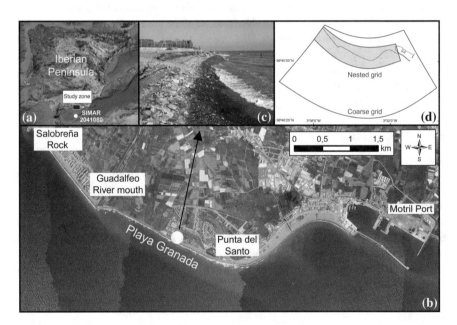

**Fig. 5.1   a** Location of the study site in the southern part of the Iberian Peninsula. **b** Aerial photograph of the study site, including the locations of the main geographical features and structures. **c** Storm erosion in Playa Granada. **d** Computational domains used in the numerical model (*Source* [31]. Reproduced with permission of Elsevier)

## 5.3  Methods

### 5.3.1  Modelled Wave Farm

A wave farm situated off the coast, in the vicinity of Punta del Santo, has been modelled in order to study the impact on wave patterns and sediment transport of wave energy extraction (Fig. 5.2). The wave farm includes eleven energy converters distributed in two rows. This layout and localization were selected based on previous works in which the numerical wave farm was optimized for coastal protection purposes [28, 29].

The wave energy converter (WEC) device chosen for the study was WaveCat [21, 22]. This is a floating WEC of overtopping type composed by two hulls Fig. 5.3 joined at the stern by a hinge [16, 20, 22]. A detailed description of this device can be found at [17, 18]. This device has been proved to be an efficient solution for the dual function of wave energy production and coastal protection ([4, 15, 28–30], among others).

The devices composing the wave farm were included as obstacles in the wave propagation numerical model using its reflection and transmission coefficients [17]. The spacing between devices was $2D$, where $D = 90$ m is the diameter of WaveCat. The baseline scenario (no wave farm) was also modelled in order to obtain the differences in wave propagation and sediment transport produced by the wave farm.

### 5.3.2  Wave and Water Level Conditions

Two sea states were modelled, representing the two prevailing wave directions at the region (westerly and easterly waves) in order to study the changes produced in the

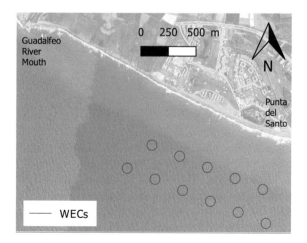

**Fig. 5.2**  Wave farm location in front of Playa Granada (*Source* [31]. Reproduced with permission of Elsevier)

**Fig. 5.3** Geometry of the
WaveCat device at a 1:30
scale (dimensions in mm)
(*Source* [31]. Reproduced
with permission of Elsevier)

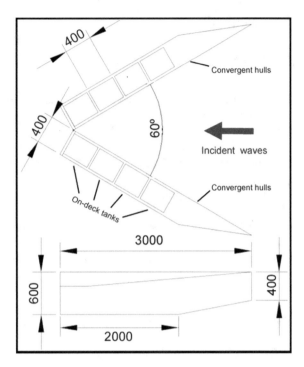

| | $H_s$ (m) | $T_p$ (s) | $\theta$ (°) |
|---|---|---|---|
| West | 3.1 | 8.4 | 238 |
| East | 3.1 | 8.4 | 107 |

**Table 5.1** Parameters of the sea states ($H_s$: significant wave height, $T_p$: peak period, $\theta$: mean wave direction)

shoreline at a storm time scale. Significant wave height and peak period were set as the most frequent values for storm conditions (Table 5.1).

As for the water level, three scenarios were studied: current situation (SLR0) and low- and high-emission projection of sea-level rise in 2100 (SLR1 and SLR2, respectively). The water level data in scenarios SLR1 and SLR2 were extracted from the representative concentration pathways (RCP) 4.5 and 8.5 obtained by [23] for the studied region.

### 5.3.3  Wave Propagation Model

A third-generation wave propagation model (Delft3D - Wave) based on SWAN model [19] was applied to obtain the influence of sea-level rise and wave farm in wave patterns. This numerical model was chosen because of the possibility of simulate the effects of obstacles on wave field as the wave energy absorption produced which means a reduction of the wave height propagating through the obstacle, the reflection of the waves that reach it, and the diffraction produced around the boundaries [24, 25, 32].

The laboratory experiments carried out by [17] were the basis to obtain transmission and reflection coefficients of WaveCat WECS, which are needed to include them as obstacles in the numerical model. Figure 5.1 depicts the two computational grid used: (i) a coarse grid which covers the deep water region, whose cell sizes vary from $170 \times 65$ m in larger depths to $80 \times 80$ m in the nearshore; and (ii) a more detailed nested grid, with smaller cell sizes ($25 \times 15$ m, approximately), which covers the wave farm area and nearshore region. This smaller cell size allows to properly reproduce the effects of each device on wave propagation.

The frequency space was divided into 37 regions logarithmically distributed, from 0.03 to 1 Hz. For the directional resolution, the circle was divided in 72 directions, in $5°$ increments, covering the $360°$. From the results of the model, which was previously validated for the study region by [12] based on the field campaigns, wave parameters at breaking were extracted and used in the LST formulation.

### 5.3.4 LST Formulation and One-Line Model

The formulation of [33] (Eq. 5.1) was applied to compute LST rates in the study region for the three sea-level rise scenarios, with and without wave farm. This formulation was developed in order to provide accurate results in different beach types as sandy, gravel and shingle beaches. Moreover, this equation has been previously applied successfully to study site [12]. The equation is as follows:

$$Q = 0.00018 K_{swell} \rho_s g^{0.5} (\tan \beta)^{0.4} (d_{50})^{-0.6} (H_{s,br})^{3.1} \sin(2\theta_{br}), \qquad (5.1)$$

where $Q$ is the volumetric LST rate, $\rho_s = 2650$ kg/m$^3$ the sediment density, $g = 9.81$ m/s$^2$ the acceleration of gravity, $d_{50} = 0.02$ m the sediment size, $\tan \beta$ the slope of the surf zone, $H_{s,br}$ the significant wave height at breaking, $\theta_{br}$ the mean wave direction at breaking and $K_{swell}$ a parameter which takes into account the effect of swell waves and varies between 1 and 1.5. This equation was applied to 341 beach profiles, evenly distributed, computing LST rates for the stretch of coast between Motril Port and Salobreña Rock (Fig. 5.1).

Shoreline position changes in each profile were tracked by means of the one-line model [27], using the LST rates obtained. Again, this model has been calibrated and validated for the study site in previous works [12]. The expression of the model is:

$$\frac{\partial y_s}{\partial t} = \frac{1}{D} \left( \frac{-\partial Q}{\partial x_s} \right), \qquad (5.2)$$

with $y_s$ and $x_s$ the shoreline position, $t$ the time, and $D$ a representative length, taken as the summation of the depth of closure and berm height.

## 5.4   Results

### 5.4.1   Wave Farm Interaction with the Wave Field

In this section, impact produced by the wave farm on significant wave height at the breaking line, $H_{s,br}$, for the sea-level rise scenarios. In order to properly compare the changes produced with respect the baseline scenario, the ratio between $H_{s,br}$ with and without wave farm will be investigated (wave height ratio from now on). In all cases, the significant wave height at breaking is reduced by the presence of the wave farm (Fig. 5.4). This reduction is less significant for the westerly storm than for the easterly storm, with ratios ranging from 0.97 to 0.98 (Fig. 5.4a), far greater than those for the easterly storm (0.79–0.8, Fig. 5.4b).

The future higher water level improves slightly the coastal defence performance of the wave farm. The largest ratio for the westerly storm corresponds to SLR0 (baseline) with 0.95, whereas SLR1 and SLR2 have ratios of 0.94 and 0.93, respectively. Moreover, with higher water levels, the area where wave power is reduced with its consequent reduction in wave height, reaches a wider coastline trench than in the current situation (Fig. 5.4). In the case of the easterly storm, there are lesser differences between the high- and low-emission scenario, with a ratio of 0.63 in both This minimum ratio rises up to 0.65 in the baseline.

### 5.4.2   LST Rate Variations

The sediment transport patterns, modified by the wave farm, are investigated in this section. For the easterly storm, the impacts on the central and western parts of Playa

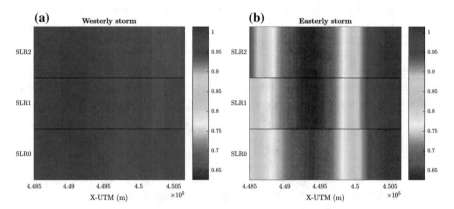

**Fig. 5.4** Ratio between the significant wave heights at breaking ($H_{s,br}$) with and without wave farm for the W (**a**) and E (**b**) storms (*Source* [31]. Reproduced with permission of Elsevier)

Granada are greater, whereas LST rates are less reduced on the eastern end of the beach (Fig. 5.5b). The shoreline horn, called Punta del Santo (Fig. 5.2), produces the observed differences in the eastern part of the beach. In the case of the westerly storm, LST rates are increased by the wave farm in the central part (Fig. 5.5a) and reduced in the eastern part of the beach.

For the comparison between the sediment transport with and without the presence of the wave farm, the ratio between LST rates with the modelled wave farm and the baseline situation (LST ratio hereinafter) has been used (Fig. 5.6). As observed before, the wave farm increases LST rates for the westerly storm in the central part of the beach, with maximum ratios of 1.53, 1.46 and 1.45 for SLR0, SLR1 and SLR2 scenarios, respectively. LST rates in the western part of the beach are decreased, conversely (ratios of 0.28, 0.29 and 0.29, Fig. 5.6a). Sea level rise produces a similar effect on LST rates than the produced for the significant wave height: it slightly increases the positive impacts produced by the wave farm. A lower alongshore-averaged ratio is observed in scenarios SLR1 and SLR2 (0.93 and 0.92, respectively) than in the baseline (0.95).

Impact produced by the presence of the wave farm is greater again in the case of the easterly storm. Greater reductions are achieved in the western and central parts of the study region, with minimum ratios of 0.26, 0.21 and 0.21 for SLR0, SLR1 and SLR2, respectively (Fig. 5.6b). Lower reductions are found in the eastern part of the beach with ratios around 1. The increased impact for the easterly storm is also depicted by the alongshore-averaged ratios for the three sea-level rise scenarios (0.51, 0.50 and 0.52, respectively).

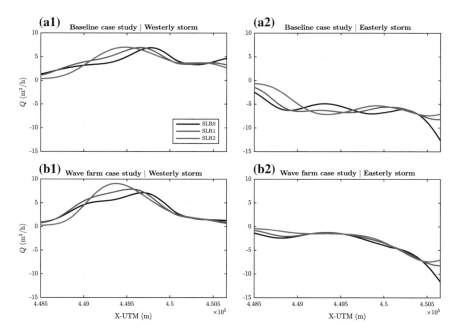

**Fig. 5.5** LST rate alongshore distribution without (**a**) and with (**b**) wave farm for the W (1) and E (2) storms (*Source* [31]. Reproduced with permission of Elsevier)

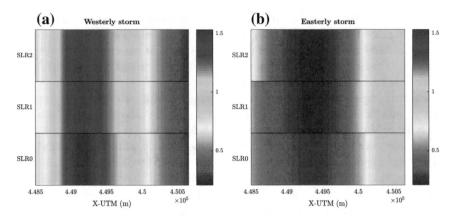

**Fig. 5.6** Ratio between the LST rates ($Q$) with and without wave farm for the W (**a**) and E (**b**) storms (*Source* [31]. Reproduced with permission of Elsevier)

### 5.4.3   Shoreline Changes

Changes in the position of the shoreline produced by the storms studied were assessed applying the one-line model by means of the LST rates described in the previous section. In order to compute these movements, the sea states were modelled for a duration of 48 h. Accretion is observed in both ends of Playa Granada for the easterly storm, whereas erosion is produced in the central part of the beach (Fig. 5.7a2). This erosion is turned into accretion with the presence of sea-level rise, more importantly in the case of the high-emission scenario. Per contra, the accretion observed in the eastern end of Playa Granada is lower with the sea-level rise.

On the contrary, erosion in the western part of the beach is observed in the case of the westerly storm, with accretion appearing in eastern end (Fig. 5.7a1). In the sea-level rise, this impact is more acute as greater erosion is observed in the western stretch of the coast and lower accretion is found in the central part of the beach, although in the east end of the beach the shoreline advance is greater. In both, easterly and westerly storm, the results around the point X-UTM = 450000 m are influenced by the derivative in Eq. 5.2 and conditioned by the changes in LST patterns.

The non-dimensional shoreline advance, indicator derived by [29], was used to assess the changes produced by the wave farm in the shoreline position. It may be expressed as follows:

$$\upsilon = \frac{\Delta y_s - \Delta y_{s0}}{\max\left(|\Delta y_{s0}|\right)}, \tag{5.3}$$

with $\Delta y_s$ and $\Delta y_{s0}$ the variation in the shoreline position with and without wave farm. Retreat or advance of the shoreline position, i.e. erosion or accretion, are indicated by negative and positive values, respectively.

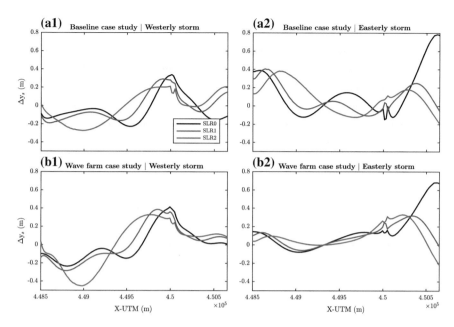

**Fig. 5.7** Shoreline advance ($\Delta y_s$) after 48 h without (**a**) and with (**b**) wave farm for the westerly (1) and easterly (2) storms. Positive (negative) values mean accretion (erosion) (*Source* [31]. Reproduced with permission of Elsevier)

The value of this indicator are shown in Fig. 5.8. Wave farm produces erosion in the western part of the beach, with accretion increasing to the east in the case of the easterly storm (Fig. 5.8b). Again, this impact is more acute with the effect of sea-level rise, whether positive or negative. Accretion is higher with the presence of sea-level rise, with maximum non-dimensional shoreline advances of 0.57 and 0.52 for scenarios SLR1 and SLR2, respectively, whereas this is 0.35 for the baseline (SLR0). Erosion is greater in the western end, reaching a minimum value of $\upsilon$ of 0.33 in SLR0, decreasing in SLR1 and SLR2 (0.69 and 0.79, respectively). The general behaviour in the whole stretch of coast, assessed by means of the alongshore-averaged values of $\upsilon$, depicts that accretion dominates due to the presence of the wave farm (0.001, 0.035 and 0.003 for SLR0, SLR1 and SLR2, respectively).

For the westerly storm, erosion is produced by the wave farm in a small stretch of coast in the western part of Playa Granada, and accretion is observed in the rest of the stretch of coast (Fig. 5.8a). The figure depicts again that sea-level rise enhance the impact produced by the farm. A greater shoreline retreat is observed, which means that erosion is higher for scenarios with sea-level rise (SLR1 and SLR2), with $\upsilon$ reaching maximum values of 0.54 and 0.57, respectively.

In the case of the baseline (SLR0), this value is lower (0.46). Accretion shows a similar behaviour with maximum values varying from 0.51 in the baseline to 0.56 and 0.61 in SLR1 and SLR2, respectively. In this case, alongshore-averaged values of $\upsilon$ are greater than for the easterly storm: 0.11, 0.10 and 0.09 for scenarios SLR0, SLR1 and SLR2, respectively.

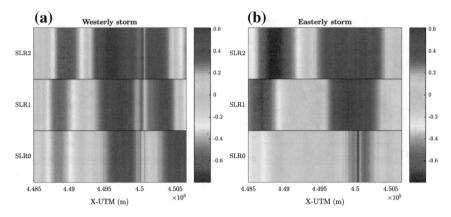

**Fig. 5.8** Non-dimensional shoreline advance ($\upsilon$) for the W (**a**) and E (**b**) storms. Positive (negative) values signify accretion (erosion) (*Source* [31]. Reproduced with permission of Elsevier)

### 5.4.4  Subaerial Beach Area Variation

Finally, the differences in dry beach area produced by the wave farm for the three sea-level rise scenarios are depicted in this section. A positive impact is produced by the wave farm under the westerly storm. Without the presence of the wave farm, erosion is found in every sea-level rise scenario, decreasing the dry beach area in 90.15 m$^2$, 42.83 m$^2$ and 51.66 m$^2$ in each scenario, respectively (Fig. 5.9a). However, with the wave farm, the erosion is changed by accretion as shown by the positive values of $\Delta A$ (2.31 m$^2$, 28.76 m$^2$ and 8.14 m$^2$, respectively). The results show that sea-level rise increase the accretion produced by the wave farm, reaching larger dry beach areas, and decreases erosion producing lower area differences.

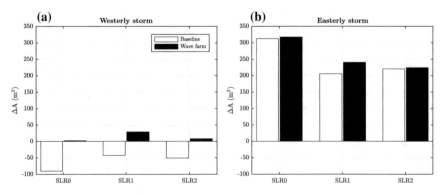

**Fig. 5.9** Subaerial beach area variation ($\Delta A$) after 48 h without (baseline) and with wave farm for the W (**a**) and E (**b**) storms (*Source* [31]. Reproduced with permission of Elsevier)

Accretion dominates in the case of the easterly storm (Fig. 5.9b), as can be observed in the results for SLR0 without the wave farm ($\Delta A = 312.6\,\mathrm{m}^2$). Sea level rise will reduce this accretion, obtaining dry beach area differences of 205.55 $\mathrm{m}^2$ and 220.38 $\mathrm{m}^2$ for low- and high-emission scenarios, respectively. However, the presence of the wave farm mitigates these negative effects, and increase dry beach area differences to 317.56 $\mathrm{m}^2$, 240.74 $\mathrm{m}^2$ and 224 $\mathrm{m}^2$ for SLR0, SLR1 and SLR2, respectively.

As the variation in dry beach area cannot be properly understood studying only the impact of sea-level rise on LST, the loss of dry beach area due to the coastal flooding produced by sea-level rise was computed. The results obtained are depicted in Fig. 5.10, which shows the total area of the beach under different sea-level rise scenarios. In the current situation (baseline), the available dry beach area is 101771 $\mathrm{m}^2$. Low- and high-emission scenarios of sea-level rise reduce this area to 88540 $\mathrm{m}^2$ and 82679 $\mathrm{m}^2$, respectively. Attending this results, 13231 $\mathrm{m}^2$ (19092 $\mathrm{m}^2$) will be lost by 2100 for the optimistic (pessimistic) projection.

For the westerly storm, final dry beach area for scenario SLR0 is decreased to 101685 $\mathrm{m}^2$, however with the presence of the wave farm this area is slightly increased to 101775 $\mathrm{m}^2$. The final area for the baseline (SLR0) in the case of easterly storm is equal to 102073 $\mathrm{m}^2$ with wave farm and 102061 $\mathrm{m}^2$ without it, confirming the positive effect. In the case of the optimistic projection (SLR1) this final area is equal to 88570 $\mathrm{m}^2$ with wave farm and 88497 $\mathrm{m}^2$ without its presence for the westerly storm, whereas in the case of the easterly waves these areas are equal to 88779 $\mathrm{m}^2$ and 88741 $\mathrm{m}^2$, respectively. Finally, dry beach area for the high-emission scenario (SLR2) for the westerly storm is 82685 $\mathrm{m}^2$ (82624 $\mathrm{m}^2$) with (without) wave farm and 82906 $\mathrm{m}^2$ (82900 $\mathrm{m}^2$) for the easterly waves.

In conclusion, the results indicate that, by 2100, between 13 and 19% of the dry beach area will be lost. However, the wave farm produces a positive effect, increasing the final dry beach surface, in all scenarios studied.


## 5.5 Discussion


Although different research works have dealt with the ability of wave farms to provide coastal protection, none of these works have studied the implications of the sea-level rise produced by climate change on the coastal defence provided by the wave farm, which is the main objective of this chapter. In the case of sandy beaches, the effects on the beach profile produced by wave farms were studied by [1–3] in a storm scale. For gravel dominated beaches, the sensitivity of the coastal protection performance of a wave farm to different parameters has been studied including the alongshore position [28] and the wave farm layout [29, 30].

This work highlights that wave farms are able to provide coastal protection even under sea-level rise. Dual wave farms—with a dual function of coastal protection and carbon-free energy generation—become a efficient strategy, combating the consequences of climate change through the decarbonisation of the energy mix

**Fig. 5.10** Initial and final subaerial beach area for the three sea-level rise scenarios without and with wave farm (*Source* [31]. Reproduced with permission of Elsevier)

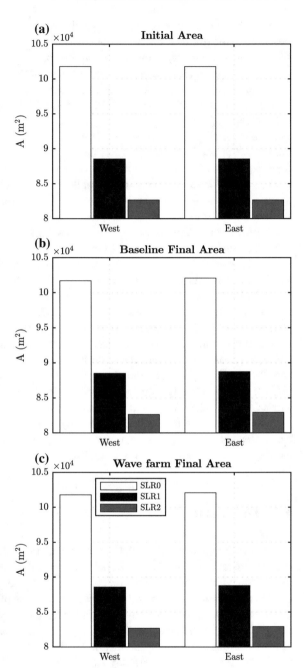

and alleviating beach erosion at the same time. On the contrary, hard-engineering solutions, such as artificial beach nourishment or groynes, are not able to provide the same level of coastal protection on the long term due to sea-level rise.

The promising results of this work notwithstanding, further research is needed in this field to properly understand the role played by wave farms in the reduction of beach erosion, passing from storm scale to a long-term scale.

## 5.6  Conclusions

Sea level rise and consequent erosion are among the repercussions produced by climate change for coastlines across the world. In this chapter, the performance of wave farms for coastal protection under sea-level rise is investigated.

For that, waves were propagated through the bathymetry using a third-generation numerical model (Delft3D-Wave), LST rates were computed by means of [33] formulation and finally the final position of the shoreline, and consequently the final dry beach surface area, were calculated using a one-line model. The results were obtained for three scenarios: the baseline (SLR0) and low- (SLR1) and high-emission (SLR2) scenarios.

Significant wave height at breaking is reduced by the wave farm in all scenarios, as depicted by the alongshore-averaged ratios with respect the current situation (without wave farm): 0.97–0.98 (0.79–0.80) for westerly (easterly) storms. The coastal protection performance of the wave farm is increased by the sea-level rise, observed as a reduction of the minimum ratios.

Consequently, this reduction in the significant wave height produce also a reduction in LST rates. The westerly storm presents alongshore-averaged ratios of 0.51–0.52 (0.92–0.95) for the easterly (westerly) storm. Again, as observed for the significant wave height, sea-level rise increases the positive effects produced by the farm, with lower alongshore-averaged LST rates, especially in the case of the westerly storm.

Finally, although some erosion is observed in the western end of the beach, this is compensated by the accretion produced in the eastern part of the beach as shown by the shoreline position obtained. The wave farm effect is confirmed positive, i.e. accretionary, taking into account the final, post storm, dry beach area. This is especially remarkable in the case of the westerly storm, where erosion is turned into accretion with the presence of the wave farm in each sea-level rise scenario studied, passing from $-90.15$ m$^2$, $-42.83$ m$^2$ and $-51.66$ m$^2$ to 2.31 m$^2$, 28.76 m$^2$ and 8.14 m$^2$ for the baseline, low and high-emission scenario, respectively. Under the easterly storm, the behaviour of the coastal system is accretionary even without the presence of the wave farm, however this accretion is larger with the modelled wave farm.

**Acknowledgements** The projects, grants, funding entities and data sources that have supported this chapter are specified in the preface of the book.

# References

1. Abanades J, Greaves D, Iglesias G (2014) Coastal defence through wave farms. Coast Eng 91:299–307
2. Abanades J, Greaves D, Iglesias G (2014) Wave farm impact on the beach profile: a case study. Coast Eng 86:36–44
3. Abanades J, Greaves D, Iglesias G (2015) Wave farm impact on beach modal state. Mar Geol 361:126–135
4. Abanades J, Flor-Blanco G, Flor G, Iglesias G (2018) Dual wave farms for energy production and coastal protection. Ocean Coast Manag 160:18–29. https://doi.org/10.1016/j.ocecoaman.2018.03.038
5. Bergillos RJ, Rodríguez-Delgado C, López-Ruiz A, Millares A, Ortega-Sánchez M, Losada MA (2015) Recent human-induced coastal changes in the Guadalfeo river deltaic system (southern Spain). In: Proceedings of the 36th IAHR-international association for hydro-environment engineering and research world congress
6. Bergillos RJ, Ortega-Sánchez M, Losada MA (2015) Foreshore evolution of a mixed sand and gravel beach: the case of Playa Granada (Southern Spain). In: Proceedings of the 8th coastal sediments. World Scientific
7. Bergillos RJ, López-Ruiz A, Ortega-Sánchez M, Masselink G, Losada MA (2016) Implications of delta retreat on wave propagation and longshore sediment transport-Guadalfeo case study (southern Spain). Mar Geol 382:1–16
8. Bergillos RJ, Rodríguez-Delgado C, Millares A, Ortega-Sánchez M, Losada MA (2016) Impact of river regulation on a mediterranean delta: assessment of managed versus unmanaged scenarios. Water Resour Res 52(7):5132–5148
9. Bergillos RJ, Masselink G, McCall RT, Ortega-Sánchez M (2016) Modelling overwash vulnerability along mixed sand-gravel coasts with XBeach-G: case study of Playa Granada, southern Spain. In: Coastal engineering proceedings, vol 1(35), p 13
10. Bergillos RJ, Ortega-Sánchez M, Masselink G, Losada MA (2016) Morpho-sedimentary dynamics of a micro-tidal mixed sand and gravel beach, Playa Granada, southern Spain. Mar Geol 379:28–38
11. Bergillos RJ, Masselink G, Ortega-Sánchez M (2017) Coupling cross-shore and longshore sediment transport to model storm response along a mixed sand-gravel coast under varying wave directions. Coast Eng 129:93–104
12. Bergillos RJ, Rodríguez-Delgado C, Ortega-Sánchez M (2017) Advances in management tools for modeling artificial nourishments in mixed beaches. J Mar Syst 172:1–13
13. Bergillos RJ, Ortega-Sánchez M (2017) Assessing and mitigating the landscape effects of river damming on the Guadalfeo River delta, southern Spain. Landscape Urban Plann 165:117–129
14. Bergillos RJ, López-Ruiz A, Principal-Gómez D, Ortega-Sánchez M (2018) An integrated methodology to forecast the efficiency of nourishment strategies in eroding deltas. Sci Total Environ 613:1175–1184
15. Bergillos RJ, Lopez-Ruiz A, Medina-Lopez E, Monino A, Ortega-Sanchez M (2018) The role of wave energy converter farms on coastal protection in eroding deltas, Guadalfeo, southern Spain. J Clean Prod 171:356–367
16. Carballo R, Iglesias G (2013) Wave farm impact based on realistic wave-WEC interaction. Energy 51:216–229
17. Fernandez H, Iglesias G, Carballo R, Castro A, Fraguela J, Taveira-Pinto F, Sanchez M (2012) The new wave energy converter WaveCat: concept and laboratory tests. Mar Struct 29:58–70
18. Fernandez H, Iglesias G, Carballo R, Castro A, Sánchez M, Taveira-Pinto F (2012) Optimization of the wavecat wave energy converter. Coast Eng Proc 1(33):5
19. Holthuijsen L, Booij N, Ris R (1993) A spectral wave model for the coastal zone. ASCE
20. Iglesias G, Carballo R, Castro A, Fraga B (2009) Development and design of the WaveCat™ energy converter. In: Coastal engineering 2008: (in 5 volumes). World Scientific, Singapore. pp 3970–3982

21. Iglesias G, López M, Carballo R, Castro A, Fraguela JA, Frigaard P (2009) Wave energy potential in Galicia (NW Spain). Renew Energy 34(11):2323–2333
22. Iglesias G, Fernándes H, Carballo R, Castro A, Taveira-Pinto F (2011) The wavecat-development of a new wave energy converter. In: World renewable energy congress-Sweden. Linköping University Electronic Press, Linköping, Sweden. pp 2151–2158. Accessed 8-13 May; 2011
23. Intergovernmental panel on climate change (2014) Climate change 2014: synthesis report. IPCC Geneva, Switzerland
24. Kieftenburg A (2001) A short overview of reflection formulations and suggestions for implementation in SWAN. Technical report, TU Delft, Department of Hydraulic Engineering
25. López-Ruiz A, Bergillos RJ, Raffo-Caballero JM, Ortega-Sánchez M (2018) Towards an optimum design of wave energy converter arrays through an integrated approach of life cycle performance and operational capacity. Appl Energy 209:20–32
26. Ortega-Sánchez M, Bergillos RJ, López-Ruiz A, Losada MA (2017) Morphodynamics of mediterranean mixed sand and gravel coasts. Springer, Berlin
27. Pelnard-Considère, R.: Essai de theorie de l'evolution des formes de rivage en plages de sable et de galets. Les Energies de la Mer: Compte Rendu Des Quatriemes Journees de L'hydraulique, Paris 13, 14 and 15 Juin 1956; Question III, rapport 1, 74-1-10 (1956)
28. Rodriguez-Delgado C, Bergillos RJ, Ortega-Sánchez M, Iglesias G (2018) Wave farm effects on the coast: the alongshore position. Sci Total Environ 640:1176–1186
29. Rodriguez-Delgado C, Bergillos RJ, Ortega-Sánchez M, Iglesias G (2018) Protection of gravel-dominated coasts through wave farms: layout and shoreline evolution. Sci Total Environ 636:1541–1552
30. Rodriguez-Delgado C, Bergillos RJ, Iglesias G (2019) Dual wave energy converter farms and coastline dynamics: the role of inter-device spacing. Sci Total Environ 646:1241–1252
31. Rodriguez-Delgado C, Bergillos RJ, Iglesias G (2019) Dual wave farms for energy production and coastal protection under sea level rise. J Clean Prod 222:364–372
32. Rusu E, Soares CG (2013) Coastal impact induced by a Pelamis wave farm operating in the Portuguese nearshore. Renew Energy 58:34–49
33. van Rijn LC (2014) A simple general expression for longshore transport of sand, gravel and shingle. Coast Eng 90:23–39

# Chapter 6
# Management of Coastal Flooding Under Climate Change Through Wave Farms

**Abstract** This chapter analyzes the effects of wave energy converter farms on storm-induced coastal flooding under three sea-level rise scenarios: present situation, optimistic projection and pessimistic projection. For that, the Delft3D-Wave and XBeach-G models were jointly apply to a gravel-dominated coast in southern Spain. The results show that the wave farm induces redutions, for the three scenarios, in breaking wave heights (about 10% and 25% under westerly and easterly storms, respectively) and total run-up values (8% and 10%, respectively). This leads to reduction in flooded cross-shore distances and dry beach areas. The decreases in flooded dry beach areas for the three sea-level rise scenarios are between 1400 and 3900 $m^2$ under south-westerly storms, and between 2100 and 3400 $m^2$ under south-easterly storms. Therefore, wave farms are efficient management strategies to mitigate coastal flooding even under sea-level rise conditions.

## 6.1 Objective

This chapter is aimed at quantifying the wave farm influence on wave propagation, total run-up and coastal flooding under storm conditions for three scenarios of sea-level rise: present situation (SLR0), optimistic projection (SLR1) and pessimistic (SLR2) projection, according to [16]. The wave farm effects were modelled through the joint application of Delft3D-Wave and XBeach-G models, detailed in Sect. 6.3, to a study area in southern Iberian Peninsula (Playa Granada, Sect. 6.2). The results were compared to those obtained without wave farm (no-farm case study).

## 6.2 Study Area

Playa Granada is located on the Mediterranean coast of southern Iberian Peninsula (Fig. 6.1a). This gravel-dominated beach, which belongs to the Guadalfeo deltaic coast, is bounded by the *Punta del Santo* horn and by the Guadalfeo River mouth to the east and to the west, respectively (Fig. 6.1b).

© The Author(s), under exclusive license to Springer Nature Switzerland AG 2020          75
R. J. Bergillos et al., *Ocean Energy and Coastal Protection*,
SpringerBriefs in Energy, https://doi.org/10.1007/978-3-030-31318-0_6

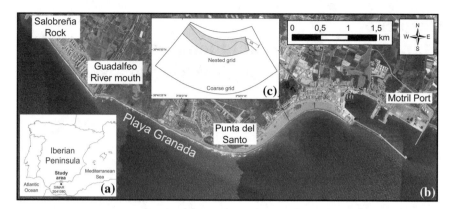

**Fig. 6.1  a–b** Location of the study area and plan view of the deltaic coast, indicating the stretch of Playa Granada. **c** Computational grid contours used in Delft3D-Wave  (*Source* [12]. Reproduced with permission of Elsevier)

The major sediment source of the study area is supplied by the Guadalfeo River [1, 2]. The river was regulated in 2004, and severe erosion and flooding issues have been frequent since then [3, 9]. The studied stretch of beach has been particularly affected (Fig. 6.2) and the nourishment projects conducted so far to face these problems have been ineffective in the medium and long term [8, 10].

This coastal region is micro-tidal (tidal range of 0.6 m) and the wave climate is bidirectional (west-southwest and east-southeast) [5, 18]. The significant wave heights in deep water with non-exceedance probabilities of 50%, 90% and 99.9% are 0.5 m, 1.2 m and 3.1 m, respectively [4]. Surge levels under storms frequently exceed 0.5 m [7].

**Fig. 6.2**  Photographs of coastal inundation events in the study area  (*Sources* [7, 21]. Reproduced with permission of Elsevier)

## 6.3  Methodology

### *6.3.1  Location and Layout of the Wave Farm*

The selected wave farm location, which is shown in Fig. 6.3, was demonstrated in previous works to be optimum for coastal protection against erosion [11, 20] and wave energy potential [17]. The farm is composed by 11 WaveCat converters, spaced 180 m and arranged in 2 rows (Fig. 6.3). This wave farm layout is also based on findings from previous works [19, 21].

### *6.3.2  Wave Conditions and Sea-Level Rise Scenarios*

The deep-water wave variables propagated with the Delft3D-Wave model towards the nearshore region, under high tide conditions, were: significant wave height of 3.1 m, spectral peak period of 8.4 s and mean wave directions of 238 ° and 107°. They are the most common values of wave variables at the study area under westerly and easterly storms, respectively. The modelled storm surge was 0.5 m, also representing a typical surge level under storm conditions.

These sea states were applied to a no-farm case study (i.e., baseline case study) and to a wave farm case study for three sea-level rise scenarios: present situation (SLR0 = 0 m) and sea-level rise values associated to the RCP4.5 (SLR1 = 0.45 m) and 8.5 (SLR2 = 0.65 m) projections estimated by [16] for the study area. They represent optimistic and pessimistic projection values in 2100, respectively.

**Fig. 6.3** Farm of wave energy converters (WECs) and beach profiles used in this chapter (*Source* [12]. Reproduced with permission of Elsevier)

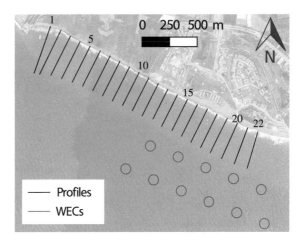

### 6.3.3  Delft3D-Wave Model

The Delft3D-Wave model, which is based on the spectral model SWAN [15], was applied to propagate the wave conditions detailed in the previous section for the three sea-level rise scenarios and the two case studies (with and without wave farm). This model was calibrated by [8] at the study area using field data measured over more than 40 days and the computational grids shown in Fig. 6.1c. These have been also used for this chapter.

The wave energy converters were introduced in Delft3D-Wave as obstacles, specifying their geometries along with their reflection and transmission coefficients. The adopted values for these coefficients are based on laboratory measurements performed by [14]. The outputs obtained with Delft3D-Wave allowed us to asses the variations in nearshore significant wave heights (Sect. 6.4.1) and to specify the input wave conditions for the application of XBeach-G (Sect. 6.3.4).

### 6.3.4  XBeach-G Model

The model XBeach-G was used to assess the total run-up values, i.e., wave run-up and water-level, for the aforementioned sea-states, sea-level rise scenarios and case studies (Sect. 6.3.2) in 22 beach profiles (one per 100 m) along the coastline of Playa Granada (Fig. 6.3). This model was calibrated at the study area by [6, 7] using field measurements collected before and after the passage of storms with varying wave conditions.

The offshore wave conditions for XBeach-G were the results of Delt3d-Wave at 10 m water depth in each profile. This offshore water depth value (10 m) fulfils the requirements detailed in manual of the model [13]. The specified land-side boundaries vary along the shoreline as a function of the type of occupation located landward of the beach profiles.

The outputs obtained with XBeach-G allowed us to quantify maximum total run-up values (Sect. 6.4.2), emerged length inundated in every beach profile (Sect. 6.4.3) and total dry beach area flooded in Playa Granada (Sect. 6.4.4) for each case study, sea-state and sea-level rise value.

## 6.4  Results

### 6.4.1  Nearshore Wave Propagation: Significant Wave Heights at Breaking

This section deals with the variations in wave propagation at the nearshore region induced by the wave farm. The ratios between significant wave heights in the wave farm case study and in the baseline case study are shown in Fig. 6.4.

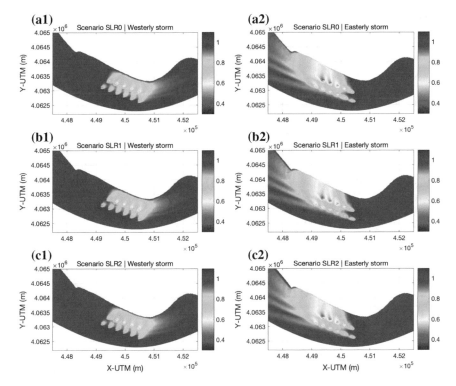

**Fig. 6.4** Significant wave height in the wave farm case study with respect to the baseline case study. **a** Scenarios SLR0, **b** scenario SLR1, **c** scenario SLR2; (1) south-westerly storm, (2) south-easterly storm (*Source* [12]. Reproduced with permission of Elsevier)

Under south-easterly storms, the wave height decreases in the lee of the farm. The reduction area comprises most of Playa Granada (Fig. 6.4a2–c2). Under south-westerly storm conditions, the wave heights are also reduced, but the decreases are lower and mainly focused on the eastern part of the beach (Fig. 6.4a1–c1). The reductions for both directions are slightly higher in absence of sea-level rise (SLR0) than in scenarios SLR1 and SLR2.

As shown in Fig. 6.5, the wave farm also generates variations in the breaking significant wave heights. Under south-westerly storm conditions, the significant wave height at breaking are reduced in the eastern part of Playa Granada up to 24.1%, 26.4% and 25.8% for SLR0, SLR1 and SLR2, respectively.

On the contrary, the breaking significant wave heights are slightly higher in the west part of the beach for the wave farm scenario (Fig. 6.5a1–b1), due to the diffraction and reflection induced by the farm. On average, the significant wave height decreases in the study area about 10% for the three scenarios.

Under south-easterly storms, the maximum reductions in breaking significant wave heights are up to 42%, 40.3% and 41.9% for scenarios SLR0, SLR1 and SLR2, respectively. They are focused in the western part of the beach, and the length

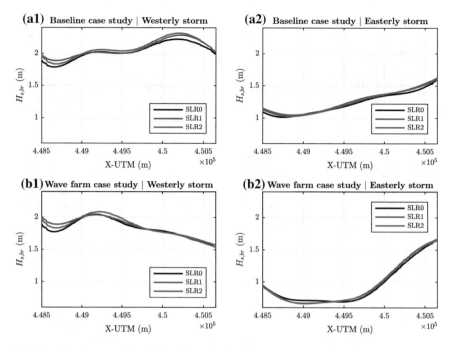

**Fig. 6.5** Alongshore variations in breaking significant wave height for scenarios SLR0, SLR1 and SLR2. **a** Baseline case study, **b** wave farm case study; (1) south-westerly storm, (2) south-easterly storm (*Source* [12]. Reproduced with permission of Elsevier)

of coastline affected by this reduction is larger than under south-westerly storms (Fig. 6.5a2–b2).

For these wave conditions, the breaking wave heights increase in the eastern part of Playa Granada due to the aforementioned reflection and diffraction processes, but theses increases are comparatively low. In this case, the average farm-induced decreases in Playa Granada are 24.8%, 25.7% and 26.3% for scenarios SLR0, SLR1 and SLR2, respectively.

In general, the significant wave heights at breaking increase with increasing values of sea-level rise, revealing that climate change do not only leads to sea-level rise but also to greater wave height values in the surf zone. This results in variations in total run-up and coastal flooding, as detailed in Sects. 6.4.2–6.4.4.

## 6.4.2  Total Run-Up

This section analyses the total run-up for all the case studies, sea-level rise scenarios and wave conditions. Figures 6.6 and 6.7 show the alongshore variations of total run-up and the ratios between total run-up with and without farm, respectively.

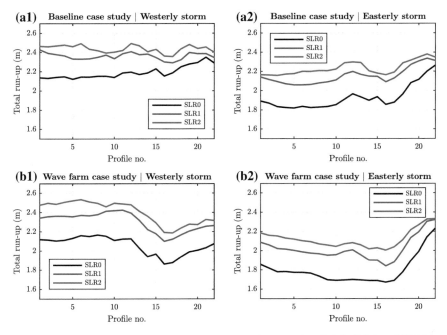

**Fig. 6.6** Alongshore variations in total run-up values for scenarios SLR0, SLR1 and SLR2. **a** Baseline case study, **b** wave farm case study; (1) south-westerly storm, (2) south-easterly storm (*Source* [12]. Reproduced with permission of Elsevier)

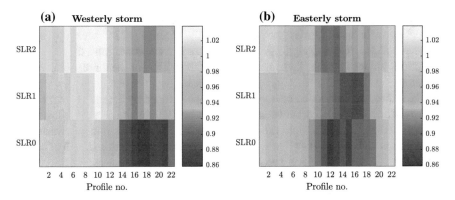

**Fig. 6.7** Total run-up values in the wave farm case study with respect to the baseline case study for scenarios SLR0, SLR1 and SLR2. **a** South-westerly storm, **b** south-easterly storm (*Source* [12]. Reproduced with permission of Elsevier)

Under south-westerly storm conditions, the wave farm reduces the total run-up more than 8% in the three sea-level rise scenarios. These reductions are mainly focused on profiles 12–22. As shown in Figs. 6.6 and 6.7, in the western part of the beach the total run-up values induced by the farm increase slightly with respect to the baseline case study. This is mainly induced by the higher significant wave heights at breaking in the wave farm case study at this location (Fig. 6.5).

Under south-easterly storms, the farm reduces the total run-up along most of the stretch of beach. Some comparatively low increases occur in the eastern part of Playa Granada (Figs. 6.6 and 6.7), but the reductions are more significant, with maximum values higher than 10% and average decreases higher than 5% for the three sea-level rise scenarios.

In general, the total run-up values under south-easterly storm conditions are lower than under south-westerly ones, influenced by shoreline orientation of Playa Granada, which is more exposed to south-westerly storms. The results of this section reveal that wave farms of WaveCat converters reduce total run-up values.

### 6.4.3   Flooded Cross-Shore Distances

The cross-shore distances inundated for each wave condition, sea-level rise scenario and case study are analysed in this section. Figure 6.8 depicts the flooded cross-shore distances, and Fig. 6.9 represents the ratios between flooded distance for the wave farm and baseline scenarios.

Under south-westerly storm conditions, the flooded cross-shore distances are decreased by the wave farm in profiles 11–22, 17–22 and 17–22 in scenarios SLR1, SLR2 and SLR3, respectively (Fig. 6.9). Under these wave conditions, the maximum reductions induced by the wave farm are higher than 10% for the three sea-level rise scenarios.

The reductions in flooded cross-shore distance induced by the farm under south-easterly storms take place between profiles 11 and 21 for the three sea-level rise scenarios (Figs. 6.7 and 6.8). In the beach profiles 1–10, the beach is overwashed in all scenarios and case studies due to the lower dry beach area in this stretch of beach, which has suffered more severe erosion problems in the past few years [2].

The maximum decreases in flooded distances due to the farm under south-easterly storms are around 12% for the three scenarios. The reductions for south-westerly storms are relevant in the eastern part of the beach, with maximum values of 7.8 m; whereas under south-easterly storms, they extend along a larger distance along the coastline, with maximum decreases of 6 m (Figs. 6.9 and 6.10).

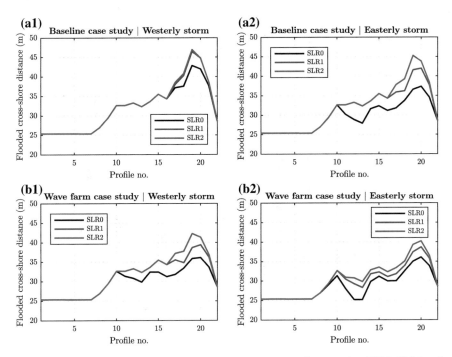

**Fig. 6.8** Alongshore variations in flooded cross-shore distances for scenarios SLR0, SLR1 and SLR2. **a** Baseline case study, **b** wave farm case study; (1) south-westerly storm, (2) south-easterly storm (*Source* [12]. Reproduced with permission of Elsevier)

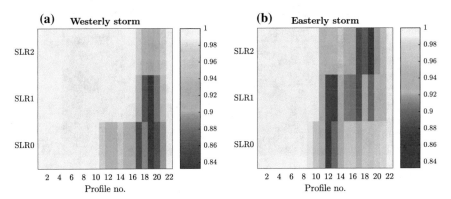

**Fig. 6.9** Flooded cross-shore distances in the wave farm case study with respect to the baseline case study for scenarios SLR0, SLR1 and SLR2. **a** South-westerly storm, **b** south-easterly storm (*Source* [12]. Reproduced with permission of Elsevier)

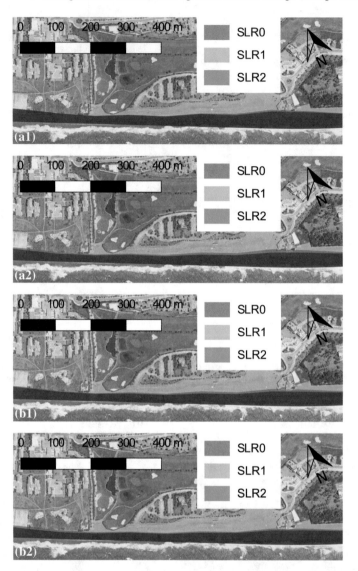

**Fig. 6.10** Dry beach surface flooded in the urbanized part of Playa Granada for scenarios SLR0, SLR1 and SLR2. **a** Baseline case study, **b** wave farm case study; (1) south-westerly storm, (2) south-easterly storm (*Source* [12]. Reproduced with permission of Elsevier)

### 6.4.4 Flooded Area

This section reports the total dry beach areas flooded for the baseline and wave farm case studies, the three sea-level rise scenarios and the two storms analysed in this chapter. They are summarized in Fig. 6.11.

Under south-westerly storm conditions, the wave farm reduces the coastal flooding in the three sea-level rise scenarios, with decreases in flooded area in scenarios SLR0, SLR1 and SLR2 about 3900 m$^2$, 2300 m$^2$ and 1400 m$^2$, respectively.

Under south-easterly storms, these reductions in flooded beach area are greater (around 2100 m$^2$, 3400 m$^2$ and 3100 m$^2$ in scenarios SLR0, SLR1 and SLR2, respectively). Under both wave directions and for the three sea-level rise scenarios, the flooded areas obtained considering the wave farm are lower than those in the baseline case study.

This highlights that wave farms contribute to reduce coastal flooding, mitigating the consequences associated to these inundation events. Thus, this management strategy is helpful to mitigate climate change effects in the coming decades.

## 6.5 Conclusions

This chapter analyses the wave farm effects on coastal flooding under storm conditions for three sea-level rise scenarios: present situation, and optimistic and pessimistic projections. To this end, Delft3D-Wave and XBeach-G were jointly applied to a study area in southern Spain to quantify the farm-induced variations, with respect to the baseline (no farm) case, in nearshore wave height, total run-up and flooded area forced by south-westerly and south-easterly storm conditions.

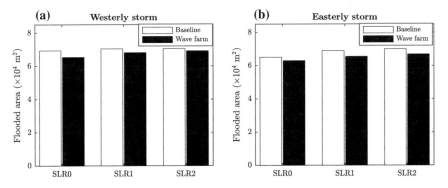

**Fig. 6.11** Dry beach surface flooded in Playa Granada for the baseline (white) and wave farm (black) case studies under scenarios SLR0, SLR1 and SLR2. **a** South-westerly storm, **b** south-easterly storm (*Source* [12]. Reproduced with permission of Elsevier)

The wave farm generates reductions in breaking wave height in the eastern part of the beach under south-westerly storms, whereas for south-easterly storms the decreases are maximum in the western part of the beach and extend over a larger distance along the shoreline.

These reductions in breaking wave height, in turn, lead to decreases in total run-up. Under south-westerly storm, the reductions are concentrated on the central and eastern parts of the beach; whereas under south-easterly storms the decreases in total run-up extend along most of coastline. The total run-up values generally greater under south-westerly storm conditions due to the shoreline orientation, which is almost perpendicular to the prevailing south-westerly storms.

The wave farm also decreases the cross-shore distances and dry beach area inundated under both storm wave directions. The reductions in flooded dry beach extension for sea-level rise scenarios SLR0, SLR1 and SLR2 are equal to 3900 m$^2$, 2300 m$^2$ and 1400 m$^2$ under south-westerly storms and equal to 2100 m$^2$, 3400 m$^2$ and 3100 m$^2$ under south-easterly storms, respectively.

According to the findings presented in this last chapter of the book, a wave farm is a management strategy not only to generate renewable energy and protect the coasts against erosion (as shown in Chap. 5), but also to reduce flooded areas under sea-level rise conditions. The results of this book will be relevant in the coming years to help mitigate coastal erosion and flooding issues associated to climate change.

**Acknowledgements** The projects, grants, funding entities and data sources that have supported this chapter are specified in the preface of the book.

# References

1. Bergillos RJ, Rodríguez-Delgado C, López-Ruiz A, Millares A, Ortega-Sánchez M, Losada MA (2015) Recent human-induced coastal changes in the Guadalfeo river deltaic system (southern Spain). In: Proceedings of the 36th IAHR-international association for hydro-environment engineering and research world congress
2. Bergillos RJ, Rodríguez-Delgado C, Millares A, Ortega-Sánchez M, Losada MA (2016) Impact of river regulation on a mediterranean delta: assessment of managed versus unmanaged scenarios. Water Resour Res 52(7):5132–5148
3. Bergillos RJ, Ortega-Sánchez M, Losada MA (2015) Foreshore evolution of a mixed sand and gravel beach: the case of Playa Granada (Southern Spain). In: Proceedings of the 8th coastal sediments. World Scientific
4. Bergillos RJ, Ortega-Sánchez M, Masselink G, Losada MA (2016) Morpho-sedimentary dynamics of a micro-tidal mixed sand and gravel beach, Playa Granada, southern Spain. Mar Geol 379:28–38
5. Bergillos RJ, López-Ruiz A, Ortega-Sánchez M, Masselink G, Losada MA (2016) Implications of delta retreat on wave propagation and longshore sediment transport-Guadalfeo case study (southern Spain). Mar Geol 382:1–16
6. Bergillos RJ, Masselink G, McCall RT, Ortega-Sánchez M (2016) Modelling overwash vulnerability along mixed sand-gravel coasts with XBeach-G: Case study of Playa Granada, southern Spain. In: Coastal engineering proceedings, vol 1(35), p 13

7. Bergillos RJ, Masselink G, Ortega-Sánchez M (2017) Coupling cross-shore and longshore sediment transport to model storm response along a mixed sand-gravel coast under varying wave directions. Coast Eng 129:93–104
8. Bergillos RJ, Rodríguez-Delgado C, Ortega-Sánchez M (2017) Advances in management tools for modeling artificial nourishments in mixed beaches. J Mar Syst 172:1–13
9. Bergillos RJ, Ortega-Sánchez M (2017) Assessing and mitigating the landscape effects of river damming on the Guadalfeo River delta, southern Spain. Landscape Urban Plann 165:117–129
10. Bergillos RJ, López-Ruiz A, Principal-Gómez D, Ortega-Sánchez M (2018) An integrated methodology to forecast the efficiency of nourishment strategies in eroding deltas. Sci Total Environ 613:1175–1184
11. Bergillos RJ, Lopez-Ruiz A, Medina-Lopez E, Monino A, Ortega-Sanchez M (2018) The role of wave energy converter farms on coastal protection in eroding deltas, Guadalfeo, southern Spain. J Clean Prod 171:356–367
12. Bergillos RJ, Rodriguez-Delgado C, Iglesias G (2019) Wave farm impacts on coastal flooding under sea-level rise: a case study in southern Spain. Science of the Total Environment 653:1522–1531
13. Deltares: XBeach-G GUI 1.0. User Manual. Delft, The Netherlands (2014)
14. Fernandez H, Iglesias G, Carballo R, Castro A, Fraguela J, Taveira-Pinto F, Sanchez M (2012) The new wave energy converter WaveCat: concept and laboratory tests. Mar Struct 29:58–70
15. Holthuijsen L, Booij N, Ris R (1993) A spectral wave model for the coastal zone. ASCE
16. Intergovernmental panel on climate change (2014) Climate change 2014: synthesis report. IPCC Geneva, Switzerland
17. López-Ruiz A, Bergillos RJ, Ortega-Sánchez M (2016) The importance of wave climate forecasting on the decision-making process for nearshore wave energy exploitation. Appl Energy 182:191–203
18. Ortega-Sánchez M, Bergillos RJ, López-Ruiz A, Losada MA (2017) Morphodynamics of mediterranean mixed sand and gravel coasts. Springer, Berlin
19. Rodriguez-Delgado C, Bergillos RJ, Ortega-Sánchez M, Iglesias G (2018) Protection of gravel-dominated coasts through wave farms: layout and shoreline evolution. Sci Total Environ 636:1541–1552
20. Rodriguez-Delgado C, Bergillos RJ, Ortega-Sánchez M, Iglesias G (2018) Wave farm effects on the coast: the alongshore position. Sci Total Environ 640:1176–1186
21. Rodriguez-Delgado C, Bergillos RJ, Iglesias G (2019) Dual wave energy converter farms and coastline dynamics: the role of inter-device spacing. Sci Total Environ 646:1241–1252

Printed in the United States
By Bookmasters